W. E. D'Arcy

Preparation of Forest Working-Plans in India

W. E. D'Arcy

Preparation of Forest Working-Plans in India

ISBN/EAN: 9783743403406

Manufactured in Europe, USA, Canada, Australia, Japa

Cover: Foto ©berggeist007 / pixelio.de

Manufactured and distributed by brebook publishing software (www.brebook.com)

W. E. D'Arcy

Preparation of Forest Working-Plans in India

PREPARATION

OF

FOREST WORKING-PLANS IN INDIA

BY

W. E. D'ARCY.

Issued from the Office of the Inspector General of Forests, Working-plans Section.

3RD EDITION.

CALCUTTA:
OFFICE OF THE SUPERINTENDENT OF GOVERNMENT PRINTING, INDIA.
1898.

INTRODUCTION.

IT is sought in these notes to explain, in a practical manner, the form in which working-plans, such as are at present required for the State forests of India, should be compiled, so that it may be possible to apply and to control them. The discussion of calculations and theories which are inapplicable in the actual condition of Indian Forestry has, as far as possible, been avoided.

The only means by which Local Governments have hitherto, as a rule, attempted to secure a supply of forest-produce for the use of the agricultural population, has been by burdening forest lands with *rights* under settlements. Hence we find, in Northern India especially, many cases in which Government has voluntarily rendered itself helpless to prevent the destruction of the forest property it desires to preserve. Such mistakes will cease to be made when it is realised that the purpose with which each forest should be managed can be prescribed by means of working-plans.

But, in order that the vast areas under the control of the Forest Department in India may be brought under the provisions of working-plans within a measurable distance of time, it is necessary that the agency of subordinate officers should be more largely utilised than has hitherto been the custom in the collection of the *data* on which these plans are based; and that this may be feasible, some such instructions as the present are obviously required.

The writer desires to acknowledge the aid he has derived from the work on *Aménagement*, recently published by Monsieur Puton, *Directeur de l'École nationale forestière* in France. He is also indebted to the courtesy of Monsieur Bartet, *Inspecteur des forêts*, in charge of the *station de recherches* at the Nancy Forest School, for copies of several working-plans in force in France. These plans, as well as several of the plans compiled by officers in India, have been largely utilised; as have also Mr. Fernandez's translation of Monsieur Broillard's *Aménagement*, and an excellent *Short Treatise on the Measurement of Timber Crops* which appeared in the *Forester* in 1889.

W. E. D'ARCY.

February 1891.

PREFACE TO THE SECOND EDITION.

These notes, which were written by the author when Assistant Inspector General of Forests and Superintendent of Working plans, were professedly intended primarily for use by subordinate Forest Officers, and have in practice amply fulfilled the object for which they were compiled. It may, indeed, be confidently stated that no professional work of the kind has had in India such widespread and beneficial results in systematising forest organisation. The scope of the book precluded the discussion of certain higher branches of forest science well known in Europe.* Moreover, in present circumstances in India, simplicity is of the first import-

---

* Those who wish to pursue the subject in all its branches may be referred to the *Manual of Forestry*, by W. Schlich, Ph. D., Bradbury Agnew & Co., London.

ance in the preparation of plans for forests which are only now being brought under systematic working; and accordingly in these notes theoretical considerations, which might in practice defeat that object, have either been omitted or been assigned a place altogether subordinate to that occupied by a brief exposition of such principles and methods of working as are more immediately applicable in this country.

The book is in much request, and the opportunity has been taken, in preparing the present edition, to revise the whole, leaving the substance as far as possible unaltered.

The suggestions contained in the work are, however, merely intended as an aid in the preparation of working-plans and are not issued as a code of official instructions.

OFFICE OF INSPECTOR GENERAL OF FORESTS,
SIMLA;
August 1895.

PREFACE TO THE THIRD EDITION.

As the present stock of D'Arcy's "Preparation of Forest Working-Plans in India" has been exhausted, a third edition has been printed to meet future requirements. The text has not, in any way, been revised, but is identical with the second edition (revised), which was published in 1895.

OFFICE OF INSPECTOR GENERAL OF FORESTS,
SIMLA;
September 1898.

TABLE OF CONTENTS.

CHAPTER I.—PRELIMINARY EXPLANATIONS.

Working-plan.

Meaning of the term "forest organisation"	1
" " "working-plan"	ib.

The Forest Capital or Producing Stock.

Meaning of the term: constitution of the producing stock	2
" " when the age-classes are irregularly distributed	4
Information afforded by the number of trees of each age-class	ib.
Chief points to be remembered with regard to the forest capital	5
Normal forest	7
Abnormal forest	ib.

The Exploitable Age.

Meaning of the term	8
Calculation of the exploitable age	9
Exploitable age for State forests	10

Increment.

Meaning of the term	12
Increments of isolated trees and crops	13

Possibility.

Meaning of the term	13
Methods of prescribing the possibility	14
Sustained yield	15

Revenue and Interest.

Meaning of the terms	15
Revenue—the net value of the produce	16
Relation of revenue to area	ib.
Interest	17

Other technical terms: definitions.

Necessity for defining such terms	17

	PAGE.
Definitions: various terms	18—22
Method of treatment, permanent methods, temporary methods, method of clearances	22
Regeneration fellings by the selection method, "storeyed" forest method, method of successive regeneration fellings	22—23
Group method, pastoral method, simple coppice, selection coppice	23—24
Coppice with standards, branch-coppice or pollard method, irregular treatment, provisional treatment, conversions	24—25
Transformations, restorations, thinnings, cleanings, weedings, working-circle	25—26
Block, compartment	26—27
Coupe	27

CHAPTER II.—PRELIMINARY WORK.

Reconnaissance of the area.

Selection of the area to be dealt with	28
Information to be collected	29
Maps	31

Choice of the method of treatment.

Sub-division of the area into portions requiring different treatment	32
Choice of the method to be applied	ib.

Formation of the working-circles.

Rules regarding the selection of circles	36

Analysis and description of the crop.

Sub-division of the area	38
Detailed description of each sub-division	ib.
The situation	39
The soil	ib.
The crop	40
Stock maps	42

Valuation surveys.

Surveys: when required; methods in use	44
Choice of a method	ib.
Method of sample plots; selection of the plots	45
Size-classes of trees enumerated	46

	PAGE
Enumeration of trees	47
Recording the results of enumeration	50

Calculating the volume of material.

Type-trees and form-factors	51

Measurement of trees and logs.

General rules	52

Calculation of the exploitable age.

General rules	58

CHAPTER III.—THE WORKING-PLAN.

Preliminary explanations.

Arrangement followed	62
Manner in which the possiblity is prescribed	ib.
General and special plans	ib.
Provisional working scheme	63
Preparatory period	ib.
Prescribing the fellings	64
Period for which fellings are prescribed	ib.
Area operated on	65
Balancing the production	66
Locating the fellings	67
Nature of the fellings	ib.
Material to be removed : possibility	ib.

Method of simple coppice.

General plan	68
Exploitable age	ib.
Period for which the fellings are prescribed	ib.
Area to be operated on	ib.
Order to be followed in the fellings	69
Nature of the fellings	ib.
Possibility	ib.
Conversion of irregular forest into coppice	71
Supplementary regulations : belts of trees	ib.
Works of improvement	72

Coppice-selection method.

General plan	72

Branch-coppice method.

	PAGE
General plan	73
Modification of the method	ib.

Method of coppice with standards.

The general plan	73
Reservation of standards	74
Supplementary regulations	76
Conversion of irregular high forest into coppice with standards	ib.
Transformation of simple coppice into coppice with standards	77

Methods of clearances.

General plan	77
Clean fellings with artificial regeneration	78

"Storeyed" forest method.

General plan	78

Selection method.

General plan	81
Limitation of the fellings : various methods	ib.
Limitation by cultural rules	82
Cases in which fellings may be limited by cultural rules	83
Limitation determined by the rate of growth	ib.
„ „ by an enumeration of the trees	84
„ „ by the productive capacity of the soil	88
„ „ by relative proportion	90
„ „ by proportionate volume	93
Length of the felling rotation	95
The fellings	96
Restoration of an incomplete crop	ib.

Method of successive regeneration fellings.

General description	97
Possibility : volumetric method	98
„ : mixed method	99
The fellings	102
Modifications of the mixed method	103
Application of the method to irregular crops	ib.

The group method.

General description	105

Pastoral treatment.

	PAGE
Area to which applicable	105
Application of method	106

Supplementary provisions of working scheme.

Subject to be dealt with	108
Improvement fellings	ib.
Sowings and plantings	ib.
Regulation of grazing	ib.
„ of rights in wood	112
Extraction of dead or fallen trees	ib.
Works of improvement other than cultural	113
Forecast of financial results	ib.

CHAPTER IV.—THE WORKING-PLAN REPORT.

General remarks	115
Arrangement of subjects	ib.

Information to be recorded under each heading.

The introduction	117
Name and situation of forest	118
Configuration of the ground	ib.
Underlying rock and soil	ib.
Climate	119
Agricultural customs: wants of the people	120
Distribution and area	ib.
Boundaries	122
Legal position of the forest	123
Rights and concessions	124
Description of the forest growth	125
Injuries to which liable	127
Systems of management	ib.
Special works of improvement	129
Revenue and expenditure	ib.
Marketable products	130
Export lines	ib.
Markets	131
Mode and cost of extraction	ib.
Net price realized for the produce	ib.
The forest staff	132
Labour supply	ib.

	PAGE
Working-circles	133
Sub-division into blocks or compartments	134
Analysis of the crop	ib.
Purpose with which forest should be managed	135
Method of treatment	ib.
The exploitable age	136
General scheme of working and calculation of the possibility	ib.
Period for which the fellings are prescribed	137
Areas to be felled annually or periodically	ib.
Rules regarding the conduct of the fellings	ib.
Tabular statement of fellings	139
Supplementary provisions	140
Improvement fellings	141
Regulation of rights and concessions	ib.
Sowings and plantings	143
Roads, buildings, and other works	ib.
Summary of works of improvement	144
Miscellaneous provisions	145
Changes in the forest staff	ib.
Forecast of financial results	146

Appendices.

Maps	146
Description of the crop in each block	147
Record of results of valuation surveys	149
Record of rate of growth	151

CHAPTER V.—CONCLUDING REMARKS.

Extent to which the provisions of a plan should be indicated on the ground by visible marks and lines 152

On the control of working-plans: its objects, and the manner in which it should be carried out . . . . 152—156

Plates.

Fig. 1. Normal forest capital of 20 years.
  „ 2. Abnormal „ „ 20 „
  „ 3. Normal „ „ 40 „
  „ 4. Sketch map, showing blocks and coupes.

PREPARATION OF FOREST WORKING-PLANS IN INDIA.

CHAPTER I.—PRELIMINARY EXPLANATIONS.

WORKING-PLAN.

Meaning of the term "forest organisation."—In order that a forest—which is here understood to mean an area set apart for the production of any forest produce—may fulfil its purpose, it is necessary that its limits and its legal status should be determined, and the object with which it is to be managed defined: that is to say, it must be demarcated, surveyed, settled, and its working prescribed in a Plan.

Although the principal product of a forest is *generally* wood in some form, this is not always so. It may be sought to obtain other products of trees, for instance, gums, resins or fodder leaves; or to preserve trees for the sake of indirect benefits derived from them, such as the protection of the soil. The only saleable product of many areas under the control of the Forest Department in India is grass; and it may be necessary, in order to obtain this, to preserve the trees, as without them the soil might become dry and barren.

All these preliminary operations, of which the preparation of the working-plan is only one, are expressed by the term *forest organization.*

The term is, however, frequently used without qualification to express the last of these operations only, namely, the preparation of the working-plan. Thus *organized forest* is generally understood to mean a forest for which a working-plan has been prescribed. An *organized crop* is one which fulfils the conditions required by a working-plan.

Meaning of the term "working-plan."—A working-plan sets forth the purpose with which a forest should be managed, so as to best meet the interests, and therefore the wishes, of the owner; and indicates the means by which this purpose may be accomplished. In other words, it is a forest regulation prescribing the application of certain cultural rules, and the execution of certain works, in order to produce a given desired result.

Trees do not usually attain the size at which they give useful products for a considerable number of years after commencing life. It follows that in a forest, worked so as to yield a regular supply, the crop has to furnish annually a product which, if it be timber at least, only matures at very much longer periods. It is generally the chief purpose of a working-plan to constitute a series of crops which will satisfy this condition.

Both the object sought and the means by which that object can be attained depend on a variety of facts relating to the forest and its management; and, in order that the prescriptions contained in the working-plan may be fully understood, it is necessary that these facts should be stated, and the manner in which the prescriptions have been deduced from them explained. These *facts*, *deductions*, and *prescriptions* are recorded in a single report which, although usually embodying several separate plans, is generally, for conciseness, called the *working-plan* of the whole area dealt with.

It is unfortunate that our term "working-plan" cannot be utilized, like its French equivalent "*projet d'aménagement*," so as to signify at once the art of preparing plans (*aménagement*), the action (*aménager*), and the person who prepares them (*aménagiste*). "Organization" has been, to a certain extent, adopted as the equivalent of *aménagement*; but, as already explained, the former term includes, as well as the preparation of the plan, the survey, settlement, and demarcation of the forest. We may, however, in the absence of a better word, agree to use *organization* in the more restricted sense. *Taxation* has been proposed, but does not appear to be generally acceptable. The French word is derived from *a* and *ménage* (household), and implies the management of a forest so as to meet the regularly recurring wants of a household; in other words, so as produce a *sustained yield*. It is, therefore, very forcible, and no good English equivalent has yet been proposed.

THE FOREST CAPITAL OR PRODUCING STOCK.

Meaning of the term.—It has been stated that the chief purpose of a working-plan is, generally, to secure the condition of crop that is necessary in order that the forest may yield perpetually a regular supply of produce in greatest quantity. It is of the utmost importance to have a clear conception of this condition, that is to say, of the constitution of the crop in an organized forest. Without such a conception it is impossible to understand the reasons for the application of even the simplest method of treatment.

In popular estimation a forest consists of a confused collection of trees of all sizes. This is not, however, the case in an organized forest capable of furnishing an uninterrupted supply of material of a definite kind. Such a forest must be built up, so to speak, of an unbroken series of trees or of crops of all ages from the seedling to the mature tree. Thus, were we dealing with isolated trees, and were it desired to fell one 20-year old tree every year, it is evident that it would be necessary to have already growing at least 20 trees, one tree 20 years old, one of 19 years, one of 18 years, and so on down to the seedling of one year. Had we not this stock of trees of different ages to deal with the desired felling could not be made. Were, for instance, all the trees of the same age, say 10 years, we should be obliged, in order to obtain one tree a year, to commence by felling a tree of that age and should end by felling a tree 30 years old.

Similar reasoning applies to *crops*, that is to say to collections of trees occupying definite areas. If it were desired to fell every year one acre of forest containing trees 20 years old, it would be necessary to possess 20 acres of forest, of which one acre should be 20 years old, one 19 years, and so on down to the crop one year old. In each case a *producing stock* of 20 trees or 20 acres is requisite in order that one tree, or one acre of forest, many annually become available for felling when aged 20 years.

The producing stock in a forest organized in this manner is part of the " capital " value of the forest ; because it acts, as will be seen, in the same way as a purely monetary capital (or what money represents) in producing *interest* in the shape of wood. In fact the forest may fairly be compared to a manufactory in which each individual tree is a workman engaged in producing wood. It must not, however, be overlooked, as bas often been done to the ruin of the forest, that the workmen in such a manufactory and the produce of their labour are identical, and that, in removing the produce, the producers may easily be removed too. The producing stock thus fulfils the economists' definition of " capital," *the produce of former labour*. The term "capital" as here used does not represent the *money value* of, or the money invested in, the forest. This includes, as a matter of course, the value of the land.

If we suppose that, as in fig. No. 1, the *growths* of the crops of different ages are represented to scale on the vertical line, and the *areas* occupied by each crop on the horizontal line, the arrangement of the crops might be graphically represented as there shown. It is evident that with a forest so constituted, there could be felled each year a crop 20 years old, one acre in extent, and that the felling might be repeated year after year. There would always be on the ground, immediately before a felling in a given year, a series of crops from 1 to 20 years old—in fact the same *wood capital*.

If, instead of this regularity, some of the age-classes were missing—if, we will suppose, there were no crops of from 10 to 18 years of age—it is evident that, 2 years after the commencement of operations, the oldest crop would be 12 years old, and it would be necessary either to fell trees of that age or wait for 8 years until the oldest crop was aged 20 years. In this case, which is graphically represented in fig. No. 2, the wood capital or stock is insufficient and is irregularly constituted, some of the age-classes being absent while others are in excess. Such irregularity in the constitution of the capital will very frequently be met with, and is, indeed, the rule rather than the exception in Indian forests.

Constitution of the stock when the age-classes are irregularly distributed.—We have hitherto assumed that the crops of different ages have been arranged in groups, following one another in regular succession, as is generally seen in the case of coppiced forests. If, however, the growths of different ages were scattered about in patches, the *capital* would still be the same, and it would still be true that, in order to fell every year an acre of forest 20 years old, there must exist on the ground a complete scale of growths, each covering one acre and aged from 1 to 20 years.

Carrying the irregularity of distribution still further we arrive at the condition of a forest such as the selection method of working leads to. The different age-classes are so intermingled that the trees overtop one another, and we practically find stems of all ages on any and every area of appreciable size. We cannot measure the area occupied by each class; but we know by analogy that the classes must exist in regular gradation of ages if the forest is to furnish uninterruptedly its full yield.

Information afforded by the number of trees of each age-class.— As regards the *number of trees* in the capital or stock of a forest, it need hardly be explained that, although the areas occupied by each age-class may be equal, the number of trees in each age-class are very far from being the same. There are infinitely more stems in a young crop, if complete, than in an older crop occupying the same area but containing trees with large crowns.

In fact, if we were actually to count the stems, from the seedling to the mature tree, on each acre in a perfectly regular forest such as that shown in fig. 1, the result, we may suppose by way of illustration, would be something as follows, though the number of stems would, of course, vary with the species and with other circumstances:—

1st	acre containing about		110 trees	20	years old.
2nd	,,	,, ,,	120 ,,	19	,,
3rd	,,	,, ,,	130 ,,	18	,,
4th	,,	,, ,,	150 ,,	17	,,
5th	,,	,, ,,	170 ,,	16	,,
6th	,,	,, ,,	190 ,,	15	,,
7th	,,	,, ,,	220 ,,	14	,,
8th	,,	,, ,,	250 ,,	13	,,
9th	,,	,, ,,	300 ,,	12	,,
10th	,,	,, ,,	350 ,,	11	,,
11th	,,	,, ,,	400 ,,	10	,,
12th	,,	,, ,,	500 ,,	9	,,
13th	,,	,, ,,	600 ,,	8	,,
14th	,,	,, ,,	800 ,,	7	,,
15th	,,	,, ,,	1,000 ,,	6	,,
16th	,,	,, ,,	2,000 ,,	5	,,
17th	,,	,, ,,	5,000 ,,	4	,,
18th	,,	,, ,,	10,000 ,,	3	,,
19th	,,	,, ,,	20,000 ,,	2	,,
20th	,,	,, ,,	40,000 ,,	1 year old.	

TOTAL 20 acres containing about 82,290 stems.

In actual practice we could not, of course, distinguish between trees differing in age by only one year. If the trees in such a forest as the above had to be counted, they would be grouped in age-classes of, say,—

I.—Stems above   20 years old ;
II.—   ,,   from   19 to 12   ,,
III.—   ,,   ,,   11 to 5   ,,
IV.—Young seedlings   below 5   ,,

and the crop would thus contain of—

Class I . . . . . . 110 stems.
  ,, II . . . . . . 1,530 ,,
  ,, III . . . . . . 5,650 ,,
  ,, IV . . . . . . abundant.

This would not indicate whether the capital was properly constituted or not. We could only say that the crop was complete, and that all stages of growth were well represented; for we cannot even approximately estimate the relative number of trees in each age-class which *ought* to be found in an irregular forest. Indeed, the relative proportion is constantly changing, and this fact is one of the chief drawbacks to the application of the selection method of working, and is the reason why there are so many different ways of attempting to calculate the possibility of such forests. Where the crops of different ages occupy distinct areas nothing is simpler than to measure these areas and to arrive at a correct estimate of the quantity of material the forest can, and ought to, produce; but, in the case of irregular crops, we can neither measure the area occupied by each age-class, nor derive by other means, from the number of trees found to be growing in the forest, absolutely accurate information as regards the sufficiency or otherwise of the capital.

Summary of the chief points to be remembered with regard to the constitution of the forest capital.—If we suppose that the darkened rectangles in figs. Nos. 1 and 2 represent the volume of material produced in each coupe in any given year, the chief facts to be remembered with regard to the constitution of the forest capital may be readily understood.

B

(*a*) In order that a forest may furnish uninterruptedly a regular supply of material of a definite kind, the capital on the ground must contain a complete series of trees or crops of all ages from the seedling to the mature tree.

This, it may be seen, is certainly true with regard to crops which are regularly constituted by area. By analogy, as has been explained, we may also assume it to be true as regards crops in which the age-classes are irregularly distributed.

(*b*) Where the part of the capital, as represented by the growing stock, is normally constituted, the material becoming exploitable in a given year is equal to the mean average production during that year over the whole area.

This is very important as showing how, by felling over yearly coupes, the annual production is obtained. It would be impossible to collect annually this production over the whole area of the forest. It will be observed from fig. 1 that the exploitable material for the year concerned (the shaded rectangle) is equal to the production during that year over all the coupes, and that this is the *average annual production*.

It follows that, in a selection-worked forest of which the capital is normally constituted, the trees that become exploitable during any period of years are equal to the average production on the whole area for that period. For instance, in a period of 10 years, it would be within the production to fell all the trees that attain exploitable size during the 10 years. The danger of applying this rule is that we cannot ascertain with accuracy whether the capital in a selection-worked forest is normal or not, and consequently are unable to say whether such a felling is or is not in excess of the production. If, for instance, all the trees were nearly mature, almost the whole of the capital might, under such a rule, be removed in one year.

(*c*) The quantity of exploitable material felled in any period bears a fixed proportion to the forest capital, but the amount of this capital varies with the age at which the forest is exploited and increases as that age is prolonged.

The truth of this most important principle will be evident from an inspection of fig. 1. We may suppose that this figure represents the wood capital on the ground. The amount of this capital remains constant, and, as will be seen, may be practically represented by one-half the rectangle, having for its base the line representing the area of the forest and for its height the growth during the exploitable age. If the exploitable age were doubled, the capital would be represented, as in fig. 3, by a similar rectangle of which the height (representing the growth) would be approximately doubled : in other words, the capital would be doubled. The produce removed would, however, remain *very nearly the same* ; because, although the coupes would be halved, their number (one for each year in the exploitable age) would be doubled.

It will be seen that the proportion between the produce felled and the capital would remain constant so long as the exploitable age remained unaltered but would vary with every change in this age.

Thus, suppose the area dealt with in the above example to be 20 acres, divided into 20 coupes of one acre each, and that the average annual production has been found to amount to 100 cubic feet solid per acre per annum. Felled at the age of 20 years, the capital on the ground would always be very nearly $\frac{20 \times 100 \times 20}{2}$ cubic feet or 20,000 cubic feet, while the quantity of material removed at each felling would amount to $1 \times 100 \times 20$ cubic feet $= 2,000$ cubic feet, or to one-tenth or 10 per cent. of the capital.

Again, suppose the age of felling to be raised to 40 years, and that the area has been consequently divided into 40 coupes containing the proper gradation of age-classes, as in fig. No. 3. The capital on the ground, even assuming the average production to be slightly higher than before, or 120 cubic feet per acre per annum, would be approximately $\frac{20 \times 120 \times 40}{2} = 48,000$ cubic feet, or more than double what it was before; while the quantity of material felled (a crop 40 years old occupying half an acre) would be $\frac{1}{2} \times 120 \times 40 = 2,400$ cubic feet (very little more than before), and would amount to one-twentieth or 5 per cent. of the capital.

Normal forest.—A forest is said to be *normal* when, in addition to being constituted of a complete series of growths of all ages from the seedling to the exploitable tree, each age-class occupying equal areas, it is completely stocked and the growth is proportionate to the fertility of the soil. It follows that there is nothing absolute in the term. A capital normal under one method of treatment or age of exploitation would be abnormal under any other age or treatment.

In a normal forest the quantity and quality of the growing stock are sufficient with regard to the age of exploitation. There are no blanks or damaged stems, the age-classes are distributed in the manner required by the method of treatment and are of proper proportion and gradation.

Abnormal forest.—A perfectly normal forest is therefore a purely ideal creation. No such forests actually exist,* although many approach the condition when they have been

* Some of the Nilgiri *Eucalyptus* plantations are almost normal.

under a regular system of organised treatment for a long period. A forest may be abnormal for one or any of the following reasons :—

The quantity of material in the capital, as represented by the growing stock, is insufficient or superabundant.

A complete scale of age-classes does not exist, and therefore the proportion between the classes is defective.

The growth is defective.

The peculiar constitution of the capital necessary in order that a forest may furnish an interrupted supply of material is sufficient of itself to explain why a working-plan is essential for the management of a forest estate, although a farm or industrial enterprise can be perfectly well managed without such a plan. The capital in the case of a forest requires many years to create, while in the case of a manufactory or of a farm the working capital, buildings, implements, seeds, live-stock, etc., can be procured in a few days or months. There are, however, many other reasons. The material produced by a forest, viz., wood, is the same as the capital or material that produces it. At the same time this material must be preserved in its growing state for years as it has no value when first produced. The forester is therefore liable to reduce his capital unawares. He is even tempted to do so, as he thereby temporarily increases his income without, perhaps, feeling the effects for many years. A farmer cannot reduce his capital without immediately experiencing the result. If he makes a mistake in his methods of farming, he can correct it immediately by the expenditure of additional money and labour. A forester can correct his mistakes, if he discovers them at all, only after a long lapse of time.

THE EXPLOITABLE AGE.

Meaning of the term.—A forest, like any other undertaking, must be managed and worked, or, as it is called, *exploited*, with a definite purpose. This purpose may of course be one of a great many, and its realisation is expressed by the term *exploitability*.

Thus it may be sought to produce material of a definite kind or size, fuel and small pieces of timber for the neighbouring population, or large trees of a certain species sâl or deodar ; or the object may be to produce the greatest quantity of material ; or again, the financial aspect alone may be considered, and it may be desired to obtain the highest annual money-revenue or the highest rate of interest on the capital (value of the soil, including the money laid out in improvements, and of the growing stock) invested in the forest. In other cases the object may simply be to preserve the trees for the sake of the shelter they afford to the soil, or in view of some other indirect benefits derived from them.

It is usual to group the various kinds of exploitability into classes, such as *qualitative, quantitative, commercial*, etc., but there is no need for these expressions. The purpose with which a forest is managed can be more easily expressed in ordinary language. Forestry could be very well conducted without employing any of the above terms.

The crop in a forest is said to be *exploitable* when it has attained the condition that is required in order to fulfil the purpose with which it is worked. The age of the trees when this condition is reached is called the *exploitable age.*

Thus, a forest exploited for quantity of fuel, or timber to furnish sleepers, would be exploitable when, the capital being properly constituted, the oldest trees were large enough to furnish in greatest quantity the fuel or sleepers of the size desired. The age of the trees, when this condition is reached, would be the exploitable age. Similarly, a forest exploited in view to the realisation of the highest rate of interest in the capital invested would be exploitable at the age when this object was attained.

The exploitable age also expresses the time required by new growth, in a crop which is being generated, to attain exploitable dimensions.

The use of the words *rotation* or *revolution* (French révolution) to express the exploitable age is due to the fact that, under the methods of treatment very generally in use *in former days*, the interval of time between successive fellings in the same area was the number of years in the exploitable age. . At present this is generally only true in the case of coppice. In most other cases the fellings pass over the area several times during the course of the so-called *rotation*. This word is therefore confusing, and will be used in these pages only to express the *felling rotation*. When referring to the age at which the trees are felled the term *exploitable age* will be employed.

The age at which trees are felled in properly managed forests is not, however, always the exploitable age. Owing to irregularities in the composition of the forest capital it frequently happens that, during a longer or shorter period which may, and strictly speaking ought to be, the same length as the exploitable age, trees or crops must be felled before or after they become technically exploitable in order that the correct proportion of age-classes may be secured.

Calculation of the exploitable age.—The price and the utility of wood depends, as a rule, on its dimensions, that is to say, on the age of the tree when felled. In order to determine the exploitable age in a forest exploited for its wood, it is sufficient for present purposes to ascertain the size of the trees that will furnish the greatest quantity of the most useful produce, *i.e.*, that will bring in the highest price or yield the highest rate of interest on the capital invested. The average age at which the trees attain this size can then be determined by observation. Where the trees are simply preserved with a view to indirect benefits, such as the shelter they afford to the soil, the exploitable age will generally approach their natural longevity.

It may be objected that the age at which a crop yields the greatest quantity of material, or the highest interest on the capital invested, is not the age at which the individual trees are the largest. This is perfectly true; but there are cases in which, if the trees were felled when the crop would yield the greatest amount of material or the highest interest, the latter of which conditions is reached at a comparatively low age, the material might not be of the most useful size or bring in most money. It would not be of much use ascertaining that, at the age of 10 years, a forest would yield the highest interest on the capital invested if the wood were under the size required, or that, at the age of 120 years, it would give the greatest quantity of material when the requirements were beams obtainable only from large trees 150 years old. Even if this were not the case, there are hardly any instances in which data are available for making such calculations. The age at which an exploitation furnishes the greatest quantity of produce is that at which the sum of the production, that is to say, the *total* quantity of material produced (including whatever has been removed or has disappeared since the young growth first sprang up) divided by the age of the crop is greatest. The material removed is of course hardly ever known, and the experiments required to ascertain the balance on the ground would, in default of suitable crops, generally be impossible in India.

Calculating the age at which the interest on the capital invested in the forest is highest, involves ascertaining by experiment the quantity of material the forest would yield if felled at various ages, and the net price at which this material could be sold; and it is, therefore, only in rare instances that the calculation can be made.

It may, therefore, be generally affirmed that, for practical purposes, the exploitable age of a forest crop is the age at which the individual trees furnish the kind of produce most wanted. This is not by any means necessarily the age at which the trees are largest: it may quite as well be the age at which the trees furnish small rafters or fuel billets.

Exploitable age for State forests.—There are two great classes of forest proprietors, the private owner and the State. With the private owner we are not here concerned. He will arrange to suit his own interests, felling the produce when it will either best meet his wants, or so as to bring in the highest annual revenue or the highest interest on the capital invested in the forest. It is, however, frequently asserted that the State, as representing the community at large, should *always* have in view the production of large trees which furnish produce most generally useful: for, while a large tree will yield small timber, large timber cannot be obtained from a small tree. This is, however, only true in so far that, as a rule, the State alone can afford to grow large trees. Private owners will not do so, as it does not pay them; so that, if large wood is required, it must be produced in forests owned by the State. The price realised is the best measure of the utility; and if large timber sells well it is certainly in demand and is useful. By felling, therefore, trees of that size for which the highest price is obtained, the public will, in most cases, though not always, be best served.

Where the price of large timber is not higher than that of small there is no pecuniary advantage in felling at an advanced age. Exclusive of very young crops and those in full decay, the *average annual* quantity of material produced is not very much altered by the age at which the crop is felled.

The following results were obtained from experiments made in a fir forest in France :—

Age of the crop when felled.	Volume of material per acre.	Average annual production per acre.
80	9,299 cubic feet	116·2 cubic feet.
90	11,441 ,, ,,	127·1 ,, ,,
100	13,481 ,, ,,	134·8 ,, ,,
110	15,300 ,, ,,	139·1 ,, ,,
120	16,755 ,, ,,	139·6 ,, ,,
130	17,963 ,, ,,	138·2 ,, ,,
140	18,785 ,, ,,	134·2 ,, ,,

Applying these results in respect of a forest, of, say, 1,000 acres (wooded area), exploited at 80 years in 80 annual coupes of 12·5 acres each, the yield would be 12·5 × 9,299 cubic feet, or 116,237 cubic feet a year; while the same forest similiarly exploited at 120 years (the age of the highest average yield) would produce 8·3 × 16,755 cubic feet = 139,066 cubic feet. The increased yield would not in all cases justify the retardation of the felling by 40 years. When, however, there is a demand for the produce, the greater size of the timber might so enormously increase its *value* as to justify felling at the more advanced age. Thus, the 116,237 cubic feet produced yearly at 80 years might be worth only four annas a cubic foot, or in all R29,000, while the large pieces, obtained from the deferred felling, might lead to an all-round rate of 6 annas a cubic foot, and so raise the price of the timber produced yearly to more than R52,000. Such examples are frequent in countries where there is an unlimited demand for the whole of the produce of the forests, but do not often occur in India.

The law of the increase in the value of the produce with its size is not always true in India, where there are few local industries and where good roads for extracting the produce rarely exist. In order that the rule may hold good there must be a sufficient demand and the means for extraction must be adequate. In the hills and out-of-the-way places where forests are found in India, it is often impossible in practice to work out large pieces of timber except at a prohibitive cost. In such cases the means of transport is the main factor in determining the size at which trees should be felled. There are even cases where, owing to the trouble and cost of converting into smaller pieces large timber which is not required, the net price of large timber is actually lower than that of small.

When, however, there is no demand for forest produce, especially for large timber, this is generally owing to want of good roads. This defect can often be remedied by a judicious outlay of funds, and should, wherever present, form the subject of careful investigation in connection with the preparation of the working-plan.

INCREMENT.

Meaning of the term.—The increase due to growth, which takes place during any given time in the volume of material in a tree or crop, has received the special name of increment. A distinction is drawn between the annual increment and the average or mean increment.

When the length of the period referred to is one year, the increase is called the *annual increment*. When the period includes a number of years, the *average* is obtained by dividing the total increase during the whole length of the period by the number of years. The result is called the *mean annual increment* or *mean increment*. The terms *annual production* and *mean annual production* are also used.

The annual increment is extremely difficult to measure or to estimate except empirically; and, as its calculation is in India of little practical importance, the methods employed to determine it elsewhere need not be here explained.

A tree 100 years old is found to contain 115 cubic feet of wood, and it is calculated that its production from the 99th to the 100th year was 1·5 cubic feet. Its annual production at 100 years is 1·5 cubic feet, while its average annual production or mean increment during its life-time at 100 years is $\frac{115}{100} = 1\cdot15$ cubic feet. Ten years later the tree is found to contain 126 cubic feet. Its average production during that period is $\frac{126-115}{10} = 1\cdot1$ cubic feet.

In the case of crops the production is necessarily expressed, with reference to the area, at so much per acre. But this cannot be done for isolated trees, because the area covered is uncertain and changes from year to year. In calculating the average production of crops, the total production during the period, including the material yielded at the passage of thinnings, etc., must be reckoned with.

Three acres of high seedling forest, 100 years old, which have already furnished in thinnings 5,000 cubic feet, are found to contain 25,000 cubic feet of timber. The average annual production or mean increment at 100 years is therefore $\frac{25,000 + 5,000}{3 \times 100} = \frac{30,000}{300} = 100$ cubic feet per acre.

13

The average annual production or mean increment during a given period, which it is sometimes of importance to ascertain, is calculated in the same way.

Increments of isolated trees and crops.—The mean annual production of an isolated tree varies greatly with the age and may be held to continually increase, while that of a crop, occupying a given area, varies much less and increases up to a certain time when it diminishes.

The following examples are taken from measurements made in India:—

Age of trees.	Isolated trees.		Age of crops.	Crops per acre.			
	Volume in cubic feet stacked.	Mean annual production in cubic feet stacked.		Volume in cubic feet stacked.	Previous yield, from thinnings, in cubic feet stacked.	Total volume in cubic feet stacked.	Mean annual production per acre in cubic feet stacked.
8	2	0·25	8	1,200	Nil.	1,200	150
10	3	0·30	10	1,600	225	1,825	182
12	7·7	0·64	12	1,725	525	2,250	187
14	12·1	0·86	14	1,810	870	2,680	191
16	19·2	1·20	16	1,910	1,172	3,082	193
18	26·1	1·45	18	2,088	1,364	3,452	192
20	33·2	1·66	20	2,324	1,494	3,818	191
22	42·4	1·93	22	2,544	1,660	4,204	191

The reason for the slight variation of the production in the case of crops is easy to understand. The quantity of material produced each year on an acre of forest, whether *covered* with young trees or old, does not differ so much as might be expected, as this production is due to the fertility of the soil, to the leaf-canopy, the action of light, etc. When the crop is young it contains, say, 40,000 or 50,000 young seedlings, each producing a very small quantity of material. When old, although each tree produces a hundred or a thousand times as much as each seedling, each acre contains comparatively few large stems. In the case of isolated trees, however, the area is not taken into account. Each tree starts as a young seedling, producing very little, and ends as a large tree producing a great deal by more or less constant rings of growth applied to an ever increasing circumference. Hence there is a constant increase in the average production of single trees.

POSSIBILITY.

Meaning of the term.—The most important calculations and provisions in a working-plan are those which relate to what is called the *possibility*.

Theoretically the possibility is the productive power of a wooded area expressed in quantity of material. Practically it is taken to mean the quantity of material which, without infringing the rules of forestry, may be felled in a forest, annually or periodically, for the time being. This latter quantity depends on the constitution of the producing stock, and on the relation which the stock bears to the age of exploitation determined upon. If the capital is sufficient and normally constituted, as in fig. 1 for instance, the possibility is equal to the average annual production over the whole area. If, on the other hand, the capital is insufficient, as in fig. 2 for instance, or superabundant, a quantity of material, less or greater as the case may be than the average increment, must be felled during a certain provisional period until the normal state is reached.

A forest of 20 acres, which it is proposed to exploit as coppice at 20 years, is found to contain a complete series of growths of all ages from 1 to 20 years, each occupying one acre. Here the possibility would evidently be a crop 20 years old growing on one acre (*vide* fig. 1). If, however, the crop were constituted as shown in fig. 2, it would be necessary, in order to obtain a sustained and annual yield of wood not less than 20 years old, to make the two crops of 19 and 20 years last for some 10 years, until the crops now 10 years old have attained or approached 20 years of age. The possibility in this case might be fixed during the provisional period as the crop growing on two-tenths of an acre per year, after which it would be in excess of the normal possibility, or certain areas might be cut over when the crop was more than 20 years old until the necessary gradation of ages (fig. 1) could be secured.

Methods of prescribing the possibility.—As it is impossible to collect the annual production all over the area, it is necessary to prescribe the realisation of the possibility in some practicable manner. The realisation of the possibility can be prescribed in three ways, *viz.*, (i) by *area*, (ii) by the *number of trees*, and (iii) by the *volume of material*. The first method is the simplest and is that followed in the case of coppice fellings, the method of clearances, as well as partly in the selection method of fellings regulated by cultural rules. The number of trees to be felled is prescribed in the case of standards over coppice and in the selection method. It is only in applying the method of successive regeneration fellings with thinnings, and, more rarely, in the selection method, that the volume of material to be felled is prescribed. In actual practice, however, in addition to prescribing the quantity of material, the areas in which the fellings are to be made are also prescribed. Practically, therefore, what is prescribed is the felling on a given area of either the crop, a stated number of trees, or a given volume of material.

But in whatever way prescribed, the realisation of the possibility should be formulated in a simple manner, easy to apply and to control, and so that it constitutes of itself the main provision of the working-plan.

A rule such as the following would fulfil this condition :—

"Each year there will be felled by the selection method in successive annual coupes of one-tenth of the total area, a number of trees not exceeding 700, in the proportion of 4 firs to 3 oaks."

Or, were the number of trees not prescribed and were the fellings limited by sylvicultural rules, and possibly, in addition, by the provision of a maximum volume of material, the rule might be to the following effect :—

"There will be felled, on one-tenth of the area each year, all dead, diseased or damaged stems, and stems that it is otherwise desirable, for sylvicultural reasons, to remove; the maximum volume of material removed being limited to 600 cubic feet on an average per acre."

A sustained yield.—As the crops we have to deal with in India are always abnormal, it is evident that the yield, although the methods devised for determining it may be the best possible, will not be exactly equal or *sustained* from period to period. Equality can no more be secured from the time when the working-plan is applied, than can the normal constitution of the crop be approximately attained without a lapse of years sufficient for its growth. Ordinarily a fairly equal yield from year to year during the length of one period may be secured; but, especially during the provisional period required for the proper constitution of the stock, even such equality may be freely sacrificed in the interest of sound or economical administration.

The German foresters always compare the yield which it is proposed to remove with the normal production or *potential possibility* as it has sometimes been called in India. Unfortunately in India, owing to only a few species irregularly distributed throughout the forest being saleable, or to the irregularity and incompleteness of the crops it is not often that the potential possibility can be ascertained with an approximation sufficient for comparison.

REVENUE AND INTEREST.

Meaning of the terms.—The income in money derived from a forest, after deducting the cost of felling and extracting the produce, may conveniently be called *the revenue*. The *net revenue* is the balance of income left after deducting all other charges, including the cost of maintaining and, if necessary, of improving the forest.

A forest yields every year 500 trees which, when converted and conveyed to the market at a cost of R2,000, sell for R10,000. The revenue amounts to R8,000. If the cost of maintaining and administering the forest, including necessary works of improvement, amounts for the year to R4,000, the net revenue is R4,000. The *gross receipts* amount to R10,000, but of this sum R2,000 must be considered to be an advance made, in order to place the produce in the purchaser's hands, which is recovered on the sale of the timber.

Revenue is the net value of produce.—It will be readily understood that, in calculations or comparisons of revenue, the revenue figures, after deducting the cost of felling and extracting the produce, must be used and not the gross receipts or total price realised for the produce. Against the latter must be debited money expended on placing the produce in the market and which in no sense represents the value of the produce in the forest. There is no case in which large timber, if taken far enough to a good market, will not realise a better price than small; but this would not make it advantageous to grow such timber if the increased cost of production and transport combined were not more than covered by the higher realisations. The *net revenue* might still be less than that realised for small material.

Statements showing the gross receipts, as entered in the accounts of the Forest Department, are too often made use of in India. How misleading these may prove will be seen from the following comparison between the true financial results of the working of the Department for the year 1888-89 and the results as they appear when the expenditure on timber works is treated as outlay on the forest and not as an advance which is recovered when the produce is sold:—

HEAD.	REVENUE.		Expenditure on the forests.	Percentage of revenue expended on the forests.
	Total.	Per square mile.		
	R	R	R	
Gross receipts as recorded in accounts	1,38,93,884	137	78,19,421	56
Revenue after deducting sums advanced for timber works	88,55,377	87	27,80,923	31

Relation of revenue to area.—For purposes of comparison the revenue may be usefully stated, in the same way as the production, with reference to the area from which it is derived.

In the example given above, suppose the area to be 4,000 acres of mixed forest, yielding teak, and worked by the selection method. The revenue per acre is $\frac{8,000}{4,000}$, or R2: the net revenue per acre is $\frac{4,000}{4,000}=$ R1.

Where the yield is realised periodically, the *average revenue* during the whole length of the period must be calculated in the same way as the average production; indeed, the one depends on the other. Where the period is long, compound interest should be included.

The average annual revenue from the Indian forests for the 10 years 1879-80 to 1888-89 amounted to R66,00,000; the average area under control during that time was about 90,000 square miles. For France, a statement prepared in 1878 showed the average revenue (net value of produce standing in the forest) to amount to 40 francs per hectare (2·47 acres) for the whole of the State forests. In the north of France, the average annual revenue was as high as 89 francs; while, in the south, in the forests open to grazing and worked on a short rotation, it was as low as 5 francs per hectare yearly.

The average annual revenue per acre derived from the Government forests of India, including of course vast areas of unworked lands, amounts therefore to about 2 annas an acre as compared with a similar revenue of R7 or R8, and even in some cases very much more, in European countries. The causes of the low revenue in India are not far to seek,—the want of demand for the produce, the ruined condition of most of the forests, the backward condition of the country, and, more perhaps than any other cause, the want of good roads and means of transporting produce from out-of-the-way places to the great centres of consumption.

Interest.—Interest, in connection with forestry, expresses the relation between (*a*) the money value of the capital invested in the forest, that is, the value of the soil of the growing stock, the total expenditure incurred on roads, buildings, settlements, surveys, etc., and (*b*) the net annual revenue derived from the forest. Problems connected with the interest obtainable under various systems of working, involving the use of mathematical formulæ, although of much theoretical interest and of practical importance in countries where forestry is in an advanced stage, have but little application as yet in India, owing to the condition of the forests, and to the want of accurate statistical *data* relating to the outturn of crops of various ages. Such problems need not, therefore, be here discussed.

OTHER TECHNICAL TERMS.

Necessity for defining terms used in forestry.—Before explaining the manner in which the different methods of forest treatment can be applied or prescribed, it is necessary to define the meaning of certain technical terms which it will be impossible to avoid using without sacrificing both accuracy and conciseness, and with regard to many of which we have, as yet, no recognised definitions in English. Correct definitions are the more necessary because, in forestry, terms are borrowed from every-day language, and their special meanings are, therefore, the more liable to be misunderstood.

Crop.—The terms *stock*, *growing stock*, *tree-growth* and *crop* all mean the collection of trees growing on a given area.

Tree.—A woody plant, having a single stem of considerable length and which is capable of attaining a height of at least 25 feet, is called a tree. The term includes arborescent palms and also bamboos; but the latter are generally spoken of as such.

Shrub—Is a plant which does not attain a height of 25 feet and which generally throws out branches at or near the ground.

Evergreen tree—Is one the leaves of which persist for at least a full year.

Deciduous tree—Is one which is leafless for some time during each year.

Conifers—Are trees which usually bear needle-like leaves.

Broad-leaved tree.—This term is used to distinguish trees which are not conifers.

Underwood.—The lowermost tier of growth in a collection of trees is called the underwood.

Brushwood.—This collective name is given to all inferior shrubs and bush-like growth. The term also includes fallen twigs and small branches in a forest.

Herbage.—Growth which does not become woody is called herbage.

Windfall—Is the term applied to trees broken or uprooted by any cause, generally by the wind.

Gregarious tree—Is one which has a tendency to form a more or less extended mass of forest composed of its own species only.

Hard-wooded trees or hardwoods—Comprise all trees of which the wood is tough and heavy. The term is used in contradistinction to *soft-wooded trees*, a name sometimes applied to trees the wood of which is considered to be of little or no value.

Bole—Means the trunk or stem of a tree, from the ground to the point where its main branches are given off.

Crown—Means the collection of main branches overtopping the bole.

Leaf canopy—Is the mass of foliage formed by the crowns of a collection of trees.

Regular and irregular forest.—The trees forming a crop may be (*a*) of the same age or size or (*b*) of different ages or sizes. In the former case, the crop is said to be *regular* or *single-aged*; in the latter, *irregular* or *of mixed-ages.*

Some recent French writers have objected to these terms *regular* and *irregular* on the ground that "regular" means "conformable to rule." The terms, however, seem to be generally accepted and understood, although "single-aged" and "mixed-aged" are to be preferred.

Forest in storeys.—Irregular crops may be composed of stems of all ages and sizes confusedly mixed, as in forests treated by the method of selection fellings; or they may be formed of trees of different ages rising above one another, as in the case of coppice with standards. In the latter case, the crop, from being arranged in tiers, is said to be in storeys.

Principal and accessory species.—In a mixed crop the *principal species* are those which, from their superior value, their abundance or the prevailing conditions for their favourable growth, determine the method of treatment applied and the age at which the crop is felled. The other valuable species, well distributed throughout the forest, are called accessory or auxiliary species.

Minor produce.—All the products of a forest, other than those of the principal or accessory species, are usually called minor produce.

Age.—The age of a regular or single-aged crop is the mean age of the trees composing it. In a crop of mixed ages, the age of the class (of trees of the principal species) most numerously represented is generally taken and alluded to as the *predominant age*. In the case of crops in storeys, the age of each tier is generally separately stated.

Pure and mixed crops.—The crop in a forest may consist of one or of several species. In the former case it is said to be a pure crop; in the latter a mixed crop.

Density.—The density of a crop is the degree of completeness of the leaf canopy of the trees that compose it. This is usually expressed by a co-efficient, but may be described by stating the species, the number of stems and their age per unit of area.

Isolated trees and canopied forest.—Crops may, from the point of view of their density, be composed either of isolated stems or of trees the crowns of which touch each other more or less so as to form a canopied forest.

Complete and incomplete crop.—In every case a crop is, as regards its density, either complete or incomplete. It is complete when it presents a density conformable to its nature and age : when this is not the case it is incomplete.

Interrupted crops.—If incomplete, a crop is said to be interrupted when it contains comparatively large blanks or gaps.

Blanks—Are areas interspersed in a forest and bare of trees.

Close, crowded and open crops.—The state of the canopied mass may also be spoken of as *crowded*, *close* or *open*, according as the lateral branches interlace or meet, or only slightly touch at their extremities when shaken by the wind.

Cover.—The term "cover" is used to express the horizontal projection of the crown on the ground, and is applied both to the surface of the ground so covered and also to the action of the cover.

Shade.—The term *shade* is applied both to the surface of the ground shaded and also to the action of the shade.

Suppression.—A plant is said to be suppressed or under suppression when its growth is injured by the cover of another.

Nurse.—Trees grown in order to protect more valuable young plants against the influences of the climate until the latter no longer require such protection are called nurses.

Seedling.—A seedling is a young plant which results directly from the germination of a seed.

High or seedling forest.—(French *futaie*, German *hoch-wald*) is a crop composed of trees which have sprung from seed.

<small>The true meaning of the French word *futaie*, of which our term " high forest " is intended to be the equivalent, is a crop composed of trees, the boles (*fûts*) of which are full grown (German *baumholtz*). This meaning is still preserved in the French term *taillis-sous-futaie*. All crops, whether derived from stool-shoots or from seedlings, that have arrived at this mature stage, are called "high-forests" in French working-plans. *Haut perchis sur souche* is a common expression in those plans.</small>

Such terms as *young seedling forest*, *mature seedling forest*, etc., are also used to express various stages of seedling forest development.

Names of crops.—Forest crops have received different names according to their stages of development.

Seedling crop.—From the germination of the seeds to the time when the newly-developed branches meet.

Thicket.—From the time of the branches meeting to the fall of the lower branches.

Pole-crop.—From the fall of the lower branches to the time when the crop attains its full height.

High forest.—From the time of the crop having attained its normal height.

Stages of growth of trees.—A separate individual is termed a *seedling* when it belongs to the first stage and the first part of the second stage; a *sapling* when it begins to lose its lower branches; a *pole* in the third stage; and a *tree* in the fourth stage.

Reserve or standard.—The trees left standing when a crop is felled, in order that they may grow to a larger size, are called standards or reserves.

Clearing.—The felling of the whole crop on a given area, whether a few standards are left or not, is called a clearing.

Clean felling.—If, however, no standards are left, a clear felling is often called a clean felling.

Stool-shoot—Is a stem springing from the stool or stump of a felled or fallen trees.

Root sucker or sucker—Is a stem spring from a root.

Coppice—Is a crop principally composed of stool-shoots or suckers. There is no terminology sanctioned by usage, similar to that employed with regard to high forest, to express the stages of development of coppice.

Regeneration fellings.—When it is sought to replace the existing by a new crop the fellings made are called principal or regeneration fellings.

Improvement fellings.—When it is sought to improve or restore the condition or constitution of an existing crop, by thinnings or weeding, the operation is called an improvement felling.

Natural regeneration.—The regeneration is said to be natural when the new crop is obtained from self-sown seedlings, from stool-shoots, or without extraneous aid from suckers.

Artificial regeneration.—When regeneration is obtained by sowings or plantings it is said to be artificial.

Names of fellings.—The principal or regeneration fellings, made in connection with each method of treatment, have received special names which will be described under that method of treatment to which they relate.

Method of treatment.—The body of sylvicultural rules regulating the manner in which a forest crop is regenerated and the produce realised is termed a method of treatment.

Permanent methods.—A method of treatment is called permanent when it is applied with a view to replace an existing by a new crop according to a definite scheme.

Temporary or provisional method.—These are applied to change the condition or constitution of an existing crop in order that it may eventually be treated by a permanent method.

Method of clearances or clearings.—This method consists in felling, in one operation on a portion of the area to be treated, the *whole* crop (or all but a few trees which are left to grow to a larger size) in such a manner as to pass over the entire area once during the time that is required for trees of the species felled to attain the exploitable size. Reproduction is obtained either naturally from seedlings already on the ground, or from seeds falling from adjacent trees, or from trees reserved for the purpose, or by artificial re-stocking.

When regeneration is obtained by natural means this method admits of several modifications. The forest may, for instance, be divided into *parallel strips*, leaving every alternate strip unfelled to afford shelter and seed until the whole of the first series has been exploited. Or the clearances may be made in regular succession over small adjacent areas and regeneration obtained by natural means. When recourse is had to artificial regeneration large areas are usually clean-felled and afterwards re-stocked artificially by sowing or planting.

The regeneration fellings made in connection with the method of clearances are called *clear fellings* or *clearings*, *strip fellings* or *clean fellings* with artificial reproduction, according to the particular modification of the method adopted.

Regeneration fellings by the selection method.—Consist in removing in a methodical manner, in accordance with sylvicultural requirements and so as not to exceed the possibility, the exploitable trees in a forest by felling them here and there, as they are found growing, either singly or in groups.

Since it would generally be impracticable to fell in one year over the entire forest, it is usual to prescribe the number of years in which the total area is to be worked, and to divide the forest into the same number of annual or periodic coupes one of which is to be exploited in regular succession each year or period.

The fellings made in connection with the method are called selection fellings.

"Storeyed" forest method.—This consists in forming a crop, according to a pre-arranged pattern, of stems of different ages the crowns of which are arranged in tiers, as is seen in the case of reserves of different ages grown in the method of coppice with standards. As in the latter method the difference in age between the trees of each consecutive tier is equal to the length of the felling rotation. The method differs from that of coppice with standards, because regeneration is obtained principally by seed instead of by stool-shoots.

Regular method or method of successive regeneration fellings.—In this method, instead of the entire crop being removed from the area exploited in a single operation, the removal takes place *gradually* in several successive regeneration fellings made from time to time as the new growth requires less and less shelter from the parent crop. In addition to this gradual exposure of the new growth, the young crop, as it grows up, is fostered by improvement fellings.

The successive regeneration feelings, made in connection with this method, are called *preparatory* or *seedling*, *secondary* and *final*. The first and second may be either *close* or *open*.

Group method.—This is merely a modification of the regular method of successive fellings and of the method of clearances by which the parent crop, instead of being removed gradually, is felled in small groups wherever patches of seedlings are found to be established. The regeneration fellings made are called *group fellings*.

Pastoral method.—This name may be applied to the treatment of areas, more or less tree-clad but of which the principal product is fodder, and in which it is necessary to preserve the trees for the sake of the shelter they afford to the soil and of the food furnished by their leaves. The treatment consists in ensuring the preservation of the trees, either by not exploiting them at all or by felling them only as they decay, or by merely lopping the branches; at the same time and in all cases limiting either the number of animals grazed or the period during which grazing is permitted, or both. Under such conditions regeneration is, as a rule, impossible, and the treatment consists in preserving the existing crop as long as possible.

Simple coppice method—Consists in clear felling a crop in such a manner that it reproduces itself naturally by means of stool-shoots and suckers. In applying this method the stumps or stools are generally felled flush with the ground, so that the shoots may in due course be established in the soil and become independent of the parent stool.

The regeneration fellings made in this method are called *coppice fellings*.

Selection coppice method.—This method (French *furetage*) consists in cutting periodically the strongest shoots, or those exceeding certain prescribed dimensions, out of each coppice clump.

The regeneration fellings in this method may be called *selection coppice fellings*; but, as the method is used in India only in the case of bamboos, they are often termed bamboo fellings.

Method of coppice with standards.—In this method, instead of felling clean the area operated on, as in simple coppice, a certain number of the most promising and valuable stems are reserved at each felling to grow to a larger size over the coppice. It is sought in this way to combine the advantages of assured regeneration, which is principally obtained, as in simple coppice, by means of stool-shoots and suckers, with the production of a certain quantity of large timber.

The regeneration fellings in this method are called *stored* or *mixed coppice fellings* when they relate to the coppice, and *standard fellings* when they relate to the trees reserved.

The branch-coppice or pollard method—Consists in lopping or pollarding the trees in rotation at regular intervals, often for cattle fodder or for manure for field crops. In some cases the trees are coppiced at a height from the ground of 4, 6, or more feet, in order to protect the young shoots from browsing animals. The fellings made may be called *loppings* or, where the whole crown is cut, *pollarding*. There is no regeneration felling properly so called. The trees replace themselves from seed haphazard, if at all, and are allowed to stand until they decay.

Irregular treatment—Forests are sometimes not subjected to any regular method of treatment, such of the produce as is saleable being realised, wherever and whenever there is a demand, by allowing purchasers to remove it under a system of permits or passes. The fellings made in such cases are called unregulated fellings.

Provisional treatment—It is often necessary in India to change the treatment which is being applied and to subject the forest to some other method. In order to do this the existing crop must be subjected for a time to special treatment so as to bring it to the condition requisite for the application of the new method. The treatment applied in such cases is called temporary or provisional.

Conversions.—Where the new method of treatment applied involves a change in the manner by which reproduction is obtained, from seed to coppice or from coppice to seed, the operation is called a conversion.

As there are three main classes of treatment, *viz.*, those in which reproduction is obtained by seed, those in which reproduction is obtained by coppice, and the mixed method of coppice with standard in which reproduction is obtained both by seed and coppice, there are six possible kinds of conversion—

High forest to simple coppice,
    „   „ to coppice with standards.
Simple coppice to coppice with standards.
    „   „ to high forest.
Coppice with standards to high forest.
    „   „   „ to simple coppice.

The fellings made in connection with the different operations required are all called conversion fellings.

Transformations.—Where it is simply sought to change one method of treatment to another of the same class, without altering the manner in which reproduction is obtained, the operation is called a transformation.

Thus it may be sought to treat as storeyed forest, or as regular high forest by the method of successive regeneration fellings, a high forest worked by the selection method, etc.

The fellings made in such cases are called transformation fellings.

Restorations.—It is frequently necessary to improve the condition of the existing crop, or to reconstitute the forest capital, by carefully respecting the main crop and by limiting operations to the cutting out of ill-grown or injured species, injurious climbers, etc , without altering the method of treatment or the mode of reproduction. The treatment temporarily applied in such cases may be called a restoration, and the fellings restoration fellings.

This operation is very common and necessary in Indian forestry, but hitherto, when styled an *improvement felling*, has not generally been carried out with a sufficiently precise purpose, i.e., with a view to the restoration of the capital according to a definite scheme and within a fixed period. The term "improvement" applied to such fellings is also too vague, as it includes all felling operations, such as thinnings and weeding, tending to improve the condition of the existing crop.

Thinnings—Is the name applied to the operation of removing the superabundant stems from an immature crop in order to give the trees of the future more light and room for growth.

Weeding.—This term is applied to fellings when carried out with the view of removing dominant inferior species. This operation frees the heads and leading shoots of the principal species in the crop.

Working-circle.—In order to arrange definitely for the working of a large forest area, it is necessary to separate those portions that require different treatment, or that it is desirable, from the nature of the demand for the produce, to exploit separately. Such sub-divisions of the area dealt with are called working-circles.

By a working-circle is thus understood *an area, subjected to one and the same method of cultural treatment, which it is determined to exploit separately under the provisions of the working-plan.* The area dealt with in a single plan of management or working-plans report may obviously include several such circles.

In each working-circle—
(1) The method of treatment is the same throughout the entire area.
(2) The boundaries should be, as far as possible, natural and not artificial.
(3) The size depends on the nature of the treatment and on administrative circumstances.*

Block.—In order to arrive at the treatment required it is generally necessary to make a complete inventory and to record the condition of the stock in the forest. To effect this with any degree of accuracy, the area, if large, must be sub-divided into smaller portions, each of which would be described separately. The sub-divisions chosen are, for convenience and simplicity, generally *natural*, that is to say, either differing from one another in situation (for instance

---

* NOTE.—The correct sub-division of a forest area into working-circles, based on sound common-sense principles, is in India more than half the plan. This has to a great extent been lost sight of in the plans hitherto made.

on the slopes of a hill-side or valley, or on opposite sides of a river) or unlike in some permanent character of the vegetation, as in the case of forests composed of different species or occupying different classes of soil, etc. The lines of demarcation of such areas thus form the skeleton, so to speak, of the forest, and are permanent land-marks. Such natural sub-divisions, generally bearing local proper names, are included in the term blocks.

It will, however, often happen in India that, owing to the area not having already been opened out, such sub-divisions cannot be formed without cutting artificial lines through the forest. In such cases the management of the forest under the working-plan will necessitate the laying out of new roads or paths for the extraction of the produce, and the lines along which it is proposed or considered likely that these new roads will proceed should be utilized as the boundaries of the blocks.

Compartment.—The crop in each block or natural sub-division would not usually be uniform as regards its composition; and where a more exact inventory of the crop is required it will be necessary to further sub-divide each block (for the purpose of making this analysis) into smaller areas, as far as possible homogeneous as regards soil and composition and age of the crop. These sub-divisions of blocks are called compartments.

Most of the working-plans hitherto prepared in India have been unnecessarily complicated by the way in which the forests have been permanently sub-divided into more or less minute compartments (so called).

Coupe.—The area felled over, or that is to be felled over in one year or period, is called a coupe. A coupe may extend over several blocks or a block may include several coupes. In all cases integral portions are very desirable. Each coupe may include several compartments.

CHAPTER II.—PRELIMINARY WORK.

RECONNAISSANCE OF THE AREA.

Selection of the area to be dealt with.—The area under forest management in a single locality generally includes separate forests requiring different cultural treatment, and, not infrequently, forests of different *classes*,—Reserved, Protected or Unclassed,—each, perhaps, demanding a different system of management. All these areas may have, however, many points in common. They ordinarily supply the same population or market with forest produce, the labourers come from the same locality, the lines of export for all are often the same, and the management and working are supervised by the same establishment. It is, therefore, generally advisable that the organisation of the whole area of Government land under forest management *in one locality* should be carried out simultaneously and be dealt with in a single working-plan report.* Generally speaking, it will be convenient to deal with the whole area of a charge, such as the Division, the Range, or, where the organisation is incomplete, with an area likely in the future to form a single range. It may be necessary in some instances to deal with the forests in a portion of a range, and, owing to special circumstances often observable in India, in some cases it may be advisable to extend the provisions of a single plan over several ranges.

There existed at one time some difference of opinion on this point in Europe. Formerly, in France, each forest estate, bearing a distinct name either because of its situation or its origin or the manner in which it was acquired by the State, was separately organised under a working-plan of its own. It has thus sometimes come about that separate working-plans have been framed for adjacent State forests in the same division or range, and, inversely, there are instances in which a single working-plan deals with forest areas in distinct administrative charges. Recently, however, the tendency in France has been to adopt the practice followed in Germany, where the whole forest area belonging to the State, or other one proprietor, included in a single executive charge (that of a *Revier förster*, *Ober förster*, *Först meister*, etc., equivalent to our Ranger) is dealt with in a single comprehensive working-plan.

In India, up to the present, no uniform course has been followed in this respect. In many cases each *class* of property (reserved or protected forest) has been separately organised under a special plan, and in some instances areas burdened with rights have even been omitted from the plan. In very few cases has what is now practically the European system of dealing with entire charges been followed.

---

* The term " working-plan," it should be remembered, is used here in the sense explained at the beginning of these Notes as meaning " working-plan report," and not the special regulation which is drawn up for each *working circle* into which the area dealt with in the report is sub-divided.

Information to be collected.—The area to be dealt with having been decided upon, the next step in the preparation of the plan is to explore the locality so as to gain a general knowledge of the configuration of the country, the distribution and composition of the forests, the marketable products they contain, the best mode of disposing of those products; as well as of all other facts likely to be useful in determining the future plan of management. The heads of examination and enquiry may be summarized as follows :—

GENERAL DESCRIPTION OF TRACTS.

Name and situation of the area dealt with.—Civil District, Forest Division, etc.

Configuration of the country.—Flat, hilly or plateau; altitude; prevailing aspects, etc.

Rock and soil.—Geological formation; underlying rock; general characters of the soil and their effect on forest growth.

Climate.—Temperature, mean and extreme, with its effect on the forest growth; rainfall, periods of drought, and rain; prevailing winds, storms; other climatic influences on forest treatment.

Population.—Numbers and condition; agricultural customs or requirements which influence the system of forest management.

THE COMPOSITION AND CONDITION OF THE FORESTS.

Distribution and area.—The distribution of the forest; total area and areas under tree-growth, grass, streams, etc., in each class of forest; enclosures within the boundaries; maps available, their degree of accuracy, etc.

Boundaries and boundary marks.—Nature and state of repair of the marks used; enclosures and their boundaries; whether a revision of the boundaries is required or not.

Legal position of the forests.—Brief history of the acquisition by Government and settlement (where there has been one) of the forest; Act and section of the Act under which gazetted, etc.

Rights.—Their origin and general character; statement of the areas burdened with rights and free from them;

summary of rights to produce; privileges or other concessions granted at the will and pleasure of Government; adequacy of the forest resources to provide for the exercise of recorded rights or of admitted privileges; actual effect of such exercise upon the crops.

General description of the forset crop.—Composition and condition of the crop; types of forest and areas occupied by each; origin of the crop; the principal species and their relative proportion and importance, habitat, mode of reproduction, size attained and rate of growth; density of the crop, blanks, and glades; state of the reproduction; principal products yielded; grass and other minor products, etc.

Injuries to which the forest is liable.—Causes of injury; fires, grazing, offences of common occurrence; the best means of regulating or combating the causes of injury.

FUTURE SYLVICULTURAL TREATMENT.

Of all the subjects to be considered that which it is desirable to bear constantly in mind when passing from crop to crop is the method of treatment which will probably be found most suitable. For a correct general apprehension in this respect, acquired sufficiently early in the enquiry, may show for instance, that the existing condition of the forest is such that its exploitation should, for the present, be very simply regulated, say by area and by a few short cultural rules only. Owing to the neglect of this precaution, work, such as the detailed enumeration of stock, which has subsequently proved wholly unnecessary for the proper attainment of the object in view, has sometimes been carried out in India, thus delaying the completion of the plan while rendering it needlessly lengthy, complicated and costly.

SYSTEM OF MANAGEMENT.

Systems of management.—Past systems of management; their defects; changes introduced and their results; management in force and its results; regulation of fires, grazing, rights, etc.

Works of improvement undertaken.—Nature of works; their object; results attained.

Revenue and expenditure.—A summary of the revenue and expenditure for the preceding 10 years to be compiled, if the information is available, for each class of forest.

UTILISATION OF PRODUCTS.

Marketable products.—Yield of timber, fuel, minor produce, etc.; quantities consumed in past years in Government works, by purchasers, by right-holders and by free grantees; value of products consumed.

Extraction of the produce.—Export roads or rivers leading from forests to centres of consumption; their adequacy and state of repair; manner and cost of extraction; road and river improvements required.

Markets.—Distance, size and importance of principal centres of consumption; produce consumed at each centre and ruling prices obtained.

Net value of produce.—Net value in the forest of each class of produce.

MISCELLANEOUS FACTS.

The forest staff.—Strength of staff and its adequacy.

Labour supply.—Abundance or reverse of the labour supply; seasons when available; cost of labour.

_{The working-plans officer will find it convenient to record the information collected on all the above points from day to day in a bound book, with a few blank pages under each heading. It will be seen that a good deal of the information required must be obtained from the existing forest records relating to previous management. The working-plans officer should go on the ground fully supplied with this information.}

Maps.—The map used in preparing the working-plan should be on a sufficiently large scale for the distinct indication of the limits of each block (or compartment where compartments have been formed) and coupe (fig. 4). A scale of 4 inches to 1 mile, in which an area of 40 acres covers one square inch of the map, is the largest scale generally required in India. Where a map on a larger scale is necessary, in order for instance to delineate graphically the situation, composition and condition of a crop which is to be subjected to special cultural treatment, an enlargement of the 4-inch map will be sufficiently accurate—quite as accurate as any description of the crop that can be made. In such a case it is not a map, in the ordinary sense, but a *picture* that is wanted. The map used should show, in addition to the boundaries and boundary marks of the forest, the natural features of the country, hills, crests of ridges, valleys, water-courses, etc.; as well as all roads, paths, fire-lines and the like. In the absence of such a map a sufficiently accurate plan may sometimes be compiled from existing village maps. In default of this it will generally be advisable, before attempting to frame a working-plan, to make a rapid survey or a sketch map of the area on such a scale and with such accuracy as may be deemed necessary.

<small>The cost of preparing maps, based on a trigonometrical survey, on the scale of 4 inches to a mile, ranges from about R50 to R100 a square mile. The money yield of even the timber-producing forests in India would not generally justify a higher expenditure on surveys. Maps on a scale of 8 inches or 12 inches to a mile, such as have sometimes been recommended, would cost, if the details were filled in with a proportionate degree of care, some hundreds of rupees per square mile. Of this fact those who recommended such maps were probably unaware. Indeed, in Europe, large scale maps are, as a rule, solely used as legal documents, in connection with the record of the boundaries and not in connection with forest exploitation.</small>

CHOICE OF THE METHOD OF TREATMENT TO BE APPLIED.

Sub-division of area into portions requiring different treatment.—It is important that the general management to be applied to each part of the forest requiring different treatment should be determined at the outset; because the nature of the subsequent operations, such as the enumeration of the stock, etc., to be carried out depends on this. The decision arrived at on this point, as well as on the connected question of working-circles can, if necessary, be afterwards rectified when the detailed examination of the forest is made.

Choice of the method to be applied.—The treatment to be adopted depends on economic and administrative as well as on sylvicultural conditions; and the choice of the treatment is generally restricted by easily ascertained facts of which the nature may be gathered from the following remarks with regard to each of the principal methods.

Non-coniferous forests, to which *the simple coppice method* is applied, can only furnish wood of small size, for the most part merely fit for fuel. Where the demand for firewood is sufficient, and other circumstances justify the application of the method, it may be adopted. It is exceedingly easy to apply and to work.

The *coppice selection method* has hitherto only been applied in India to bamboo forests. There do not appear to be any good grounds for extending its application to other kinds of crops.

The *branch coppice* method must not, of course, be applied in areas set aside for the production of timber; but it is useful under certain circumstances, for instance where fodder is more valuable than timber, and where a regular supply of small fuel and leaf fodder must be furnished by forests worked for local use. The method may also in some cases be employed in connection with the management of pasture lands.

The *method of coppice with standards* is admirably adapted to the circumstances prevailing very generally in the plains of India, and ought to be more largely made use of than is actually the case. It meets, generally better than any other method, domestic requirements in small fuel, and at the same time furnishes a considerable quantity of timber of large dimensions suitable for the manufacture of implements and furniture. As compared with tree-growth in high forest, the reserved stems increase in girth more rapidly; and many species essentially light-loving accommodate themselves readily to this treatment. In a country where a fluctuating demand is the rule, the method has also this advantage that the standard trees can be felled in greater or lesser number as is required, or may be allowed to grow to a larger size without disorganising the working. The working-plan itself is, moreover, exceedingly simple in its arrangement, easy both to understand and to apply. As regeneration is principally obtained by means of coppice, the method can only be applied to broad-leaved species. The whole of the produce must, as a rule, be saleable to make the application of the method profitable.

* *Toungya* cultivation, it need hardly be said, should

---

* Teak *toungya* cultivation in Burma means a combination of arboricultural operations with shifting cultivation as practised by wild tribes, who cut and burn the existing vegetation in order to raise one or two crops of cereals or other food crops or cotton. When these crops are sown, small teak seedlings are planted by the cultivators at the same time, usually at a distance of six feet by six feet. They are carefully weeded for a few years; and the result is that a much more valuable tree-crop springs up than that which originally had possession of the ground.

Cultivation of this kind, but in which the subsequent artificial rearing of young tree is generally wanting, is called *jhúm* in Bengal and Assam; *khil* and *korali* in the North-West Himalaya; *bewar* in the Central Provinces; and *kumri*, *podu*, etc., in South India. Similar cultivation is or was practised in some European countries.

only be permitted where there are forest tribes who live by *toungya* or *jhum* clearing. It is rather a manner of organising the cultivation of cereal crops, so as to do the least harm to the forest, than a method of forest treatment; but it can be turned to useful account in enriching the forests by having the areas planted up with valuable species of trees where the necessity for such cultivation exists. It may possibly be hereafter combined with forest treatment so as to form a regular method.

Where the whole crop is saleable, and in climates where natural reproduction is assured, the *method of clearing* or *clean felling adjacent areas* may be applied in certain cases for the sake of its extreme simplicity and the order it introduces in the fellings. It should not, however, be made use of where the working-circle and, consequently the *coupes* are very large; as it leaves reproduction to chance, and the larger the coupes the smaller this chance.

Like the preceding, the *method of clean felling with artificial reproduction* requires the whole crop to be saleable. The climate should also be one in which artificial re-stocking can be undertaken with certainty of success, and labour should be abundant and cheap. The application of the method should not, as a rule, be attempted under other circumstances or where reproduction can be secured by natural means.

As in other clearance methods, *the strip method of clearances* can only be applied where the entire crop is saleable. A special objection to the strip method is, however, the cost of laying out the position of the fellings on the ground. Unless this is well done and the strips are permanently marked off in advance there is apt to be confusion; and in any case the work demands a good deal of time and attention from the staff. The method is unsuited to badly stocked areas and hilly or broken ground. It has the further objection of extending the felling work over a larger area than would be the case under other methods.

The "*storeyed*" *forest* method is adapted to coniferous tracts where the coppice method cannot be applied or where the uncovering of the soil would foster a dense undergrowth. It is exact and well-defined, and it leads by an easily understood system to the correct constitution of the capital. It also has the advantage of furnishing wood of all classes and sizes. It is a question, however, whether the method could be adapted to a mixed crop in which only one or two species are saleable.

The *selection method* of working, formerly represented as barbarous and unscientific, is the only method which can at present be applied in exploiting the great majority of our Indian mixed forests in which only a few species are saleable. The prescriptions necessary in order to apply it are simple, and it is well suited to the restoration, without the aid of expensive works of artificial reboisement, of the irregular and ruined forests so frequently met with in India. It adapts itself to almost any system of culture and to the special requirements of any crop or species. For instance, the cover can be removed from the seedlings by successive fellings undertaken gradually or at once, as in the regular method or in the group method. Its drawbacks are first, that, as only a portion of the material standing in the coupe is felled, a large area has to be worked over at each operation, and the extraction of the produce is expensive; but this must always be the case where only some of the component species of a crop are saleable.

On the other hand the sylviculturist can do but little towards fostering the growth; the timber is of poorer quality than that grown on areas treated by the regular method; much damage to young growth is often caused by the fellings; the accurate calculation of the possibility is impossible, and in India the danger from fire is greater.

A forest, treated by the *method of successive fellings*, should contain in compact blocks a properly graduated series of crops of different ages, mature high forest, pole crops, tickets, etc., each class occupying nearly the same extent of ground. Such a condition does not at present exist in any forest in India, and can only be induced by subjecting the area to transformation fellings during a lengthy period. But the chief drawbacks in India to the employment of the method are that it is complicated and that to apply it successfully a staff of thoroughly competent foresters is required. Different classes of fellings, each requiring to be executed with skill, must be carried out simultaneously in different parts of the area treated: principal regeneration fellings in one locality, selection fellings in another, weedings in a third, and thinnings in a fourth. It also fails in one of the chief objects of a working-plan, in that it does not tend (unless the modified method is adopted) to introduce order into the fellings. The regeneration fellings are prescribed by volume of material, a process which alone involves a laborious and costly calculation of the contents of the crop extending over a large area, and the fellings are carried out in this area wherever required. The method can only be employed where all the produce of the forest is saleable, and it is also unsuitable where reproduction is very slow or difficult to obtain, as at high elevations and in very dry climates. In fairly moist climates in the plains it might be applied where the demand for the produce is sufficiently great and profitable.

The above remarks apply also to the modification of the method of successive fellings known as *group method*.

It will be seen that in determining the method of treatment the chief points to be noticed are :—
(1) The produce in demand.
(2) The composition, condition and sylvicultural requirements of the crop.

FORMATION OF THE WORKING-CIRCLES.

Rules regarding the selection of circles.—The reconnaissance completed and the general method of treatment adapted to each section of the forest area determined, the working-circles, subject to such modifications as a more detailed examination of the crops may subsequently prove necessary, are decided upon. The most important rule to bear in mind is that all the crops included in one working-circle must be susceptible of the same method of treatment.

Each working-circle forms a separate centre of supply and a unit of administration, and may, where possible, constitute a separate charge. Very often, as for instance where there are large forest masses—such as the teak forests of Burma or the deodar forests of the Himalayas—subjected to the same method of treatment, the formation of working-circles depends entirely on administrative facts and presents no difficulty. But where there is a local demand for the produce, requiring a continuous supply near at hand in several different centres of consumption, or where the nature of the crop varies very much from place to place, it will be necessary to form a number of comparatively small working-circles each of which cannot constitute a separate charge. This is especially the case with regard to coppice forests, the principal product of which, fuel, cannot, owing to the expense, be carried to a long distance, or where, as in the Thana forests of Bombay and in the Central Provinces, the people *must* be provided on a great scale with fire-wood and small building timber. Local grazing rights or privileges also play an important part in determining the areas to be formed into separate circles; as the closings and openings of blocks to grazing often determine the cultural treatment.

Whatever boundaries are adopted should, as far as possible, be natural features and not artificial lines cut through the forest. Water-partings (in the hills), roads, boundaries of forests or of forest blocks form the most convenient limits of working-circles from all points of view.

Where the *size* of working-circles cannot be determined by the circumstances already described, no very definite rules can be laid down. If small, the number of separate circles, and consequently of separate fellings or other operations, becomes inconveniently great, and the work of a given year is correspondingly scattered. This may be objectionable from an administrative point of view. The size is obviously closely connected with the average area of the annual coupes, which in its turn usually depends on the demand for the produce. As a general rule, more produce than is saleable should not be felled, though this consideration may be sometimes disregarded when it is sought to *improve* the growth or to *restore* the forest to its original condition. If the circles are too large, the area to be exploited in one place or at each operation may be inconveniently extensive, and the distance to which the material must be conveyed too great; or more produce may have to be felled than can be consumed in the centre of consumption to which it must be transported if it is to be utilised at all. A middle course should, therefore, be chosen. Generally speaking, where all the produce of the fellings is saleable, the working-circles would be comparatively small. Where, however, only one or two species, which grow scattered in a mixed crop are exploited, the size of the working-circles is necessarily very large.

It should be remembered that where equality of yield from year to year is desired, such equality can be better secured by forming a number of small, rather than a few large, circles.

Although each working-circle is subjected to a single

series of principal fellings, one or more separate series of minor operations in connection with the method of treatment adopted may be carried out in the same working-circle.

For instance, in a circle worked by the method of successive fellings and thinnings, provisional selection fellings might be carried on in one block, thinnings in another, and so on. But in this case the principal or regeneration fellings would, in time, pass over the whole area, and the minor operations would have the common object of leading on the crops until they reached maturity and could be regenerated in turn by successive fellings under one and the same method of treatment. In some cases, as for instance in exploiting bamboos or cutch trees in the teak forests in Burma, a series of operations under one plan of work (for cutch) may have to be carried out in a number of working-circles overlying those formed under other plans (for teak).

ANALYSIS AND DESCRIPTION OF THE CROP.

Sub-division of the area.—In order to describe, which is the next step, the composition and condition of the forest crop, it is necessary to sub-divide the area of each working-circle into smaller areas or blocks bounded by natural limits, and possibly into compartments. The size of the blocks is important; but no precise rules can be laid down with regard to this subject which depends on the nature of the work and on the minuteness indicated by the treatment. To a certain extent the size will be of course decided by existing facts. The different portions of the forest having natural limits vary in extent, and the various kinds of crops may be found over large areas or small. But by grouping or by sub-dividing, as the case may be, the existing divisions extremes can always be avoided. If the divisions are small they may be too numerous, and the result is confusion in the plan and a tedious number of separate descriptions. On the other hand if too large, the inventory of the forest is vague and unsatisfactory.

Detailed description of each sub-division.—The work of describing the areas into which the forest is sub-divided should proceed simultaneously with the sub-division itself. The minuteness of the description depends, as does the sub-division of the area, on the method of treatment adopted and on the object in view.

The most desirable record may, therefore, vary from a broad general indication of the state of the crop in each block, accompanied by the results of such enumeration surveys as have been carried out, to a detailed and separate account of the situation, and of the soil and stock found in each compartment into which the blocks have been differentiated.

<small>The work requires trained observation a nd a good deal of physical exercise but does not call for any involved process of reasoning. It is necessary, however, that the attention of working-plans officer should be continuously sustained so that no important facts may escape him. He should not, therefore, attempt more at one time than can be accomplished without over-fatigue.</small>

The situation.—The situation includes the relative position and elevation as well as the aspect and slope. As regards *elevation*, the absolute height above sea-level should be noted generally for the forest; but the *height relative* to the surrounding country, together with the absence or presence of sheltering land, is of more importance as regards particular blocks or compartments and should be noted. Thus the upper portion of a slope near the top of a ridge may require very different cultural treatment from the lower portion towards the bottom of the underlying valley, although the difference of level above the sea may be slight. The *aspect* should always be stated where it is well defined; but in hilly ground a single block often faces several points of the compass. The *slope* may be stated in a single word. A slope is said to be gentle when the inclination is not greater than about 1 in 6; it is steep when more than 1 in 6 but not greater than 2 in 3; it is very steep when more than 2 in 3, and becomes precipitous when it reaches 1 in 1.

The soil.—The more important facts with regard to the *soil* may be expressed in a few words descriptive of its surface, its composition, its physical state, its depth and its fertility. The fertility or productive value of the soil, as regards the species which has to be considered, may generally be summed up by such terms as "good" or "very good," "bad" or "very bad," as the case may be. With regard to its *surface*, the soil may be quite bare and hard, or covered with a layer of leaves, or with vegetable mould apt for the reception of seed. Or it may be carpeted with moss or grass or over-grown with bushes. The physical character of the soil—its looseness and the size of the particles forming it—are, however, of greater importance as regards forest vegetation than the chemical composition. The soil may be formed of stiff clay, loose sand or agglomerations of stones and boulders; while its hygroscopicity may vary from marshy to dry. But of all the properties of the soil *depth* is perhaps the most important and is that which is most likely to vary. Depth of soil is at once manifested by the appearance of the trees. If shallow, the boles are generally short and the crowns low; while the contrary is the case in soils of from 12 to 20 inches and over in depth.

The nature of the underlying rock and sub-soil is often of great importance.

Very little is known as yet of the influence exerted by different soils on the growth and production of the various species of Indian forest trees: it is not often, therefore, that a description of the chemical characters of the soil can be of much service. But when a detailed plan is prepared, the nature of the soil in each compartment should be noted, and the relative proportions of clay, sand and lime should be roughly ascertained. By noting such particulars many valuable indications can be deduced, especially when a complete record of the production has been kept for a number of years. Small portions of the soil it is wished to examine should be taken, from *several different spots* in the nursery or compartment under observation, and should be mixed together. A small portion should then be reduced to powder in a mortar and the vegetable detritus removed by heating. The residue should then be treated with dilute hydrochloric acid, and the insoluble portion, consisting of clay and sand, separated by filteration from the soluble portion, (calcium, iron, etc.). The clay can then be separated from the heavier sand by decanting several times with the aid of a syphon, and the relative proportion of each determined by weighing. If, when the hydrochloric acid is first added, there is a *brisk* effervescence, the soil is calcareous. The quantity of lime present can be roughly ascertained by deducting the weight of the insoluble clay and sand from the weight of the mass before it was treated with the acid.

The crop.—The composition and condition of the standing crop must necessarily be examined and recorded with the greatest care. The detail depends, as already explained, on various circumstances; but, where a complete inventory is required, the following points should be considered, *viz.*:—

(1) General character of the crop.
(2) Component species, or different types of crop, and their relative proportions.
(3) Age; density; state of growth; proportion of unsound trees.
(4) Natural reproduction; presence or absence of seedlings or stool shoots.
(5) Origin and past management.
(6) Most suitable future treatment.

The character of the crop.—Where it has been found necessary to divide each block into fairly homogeneous units or compartments the character of the growth should be explained by its descriptive name in a single word or term, such as "thicket," "irregular mature seeding forest." Where, however, the sub-divisions are not artificial and the crop is irregular, a more lengthy description becomes necessary.

With respect to the *component species of trees*, the principal, and sometimes the secondary species, as well as where necessary their numbers or relative proportions, need alone be mentioned; but this should always be done in the most simple manner possible. Frequently it is sufficient to state whether species other than principal are abundant or rare. Numerical ratios may be misleading, for instance, where there are a large number of injured or unsound trees, or where the species are very irregularly distributed. Species of minor importance need only be mentioned collectively if they are abundant or rare, or are characteristic of the soil or state of growth, or if they are confined to certain compartments. Where there is more than one stage of growth, as in coppice with standards, each stage should be separately described.

As regards *age*, if the crop is composed of two or more distinct classes, the respective ages of each should be given. If the crop is irregular, the *dominant age* should be stated.

The *density* of a crop is described by stating whether the leaf canopy is complete or not, *close, open, varying*, etc.

The state of growth can generally be explained in a few words, such as "active" or "slow;" but the *condition* of the crop, and such facts as the proportion of unsound trees, the probable time during which the trees will continue in good condition, etc., sometimes require to be specially explained.

It may occasionally be possible to note the *origin* of the crop. As a rule forest history cannot be traced very far back; but in any case it is expedient to mention what operations have been last carried out. Where fires, grazing, etc., have caused serious damage which has left its traces in the crop, this should be stated and the effect noted. Generally, there is peculiar to each crop, as to each forest, some special characteristic which, if not observed and noted, will render the descriptions of compartments, however lengthy and carefully drawn otherwise, deceptive.

Lastly, it is useful to note the probable future treatment or the cultural operations that might usefully be executed in the immediate future. It does not necessarily follow that such operations will be carried out. This will depend on various considerations which will be discussed in connection with the organisation of a forest considered as a whole. But such notes are of great use when the time comes for preparing the plan.

As an example of a description of a block, the following is given :—

Name.—Dalnur.
Area.—313 acres, of which 6 acres blank, and 10 acres unproductive rock.

Situation.—On the western flank of the great Maura ridge. Aspects generally westerly, but on minor spurs some areas face to north and south. Gradients steep, in places precipitous. Elevation from 6,500 to 8,000 feet; sheltered from east.

Soil.—Rich loam well covered with mould; generally deep and suited to deodar, except near summit where underlying rock of black limestone crops out.

Stock.—Irregular mixed crop. In the upper part, *kharsu* oak with a few blue pine, spruce, maples, and rarely deodar. Lower down deodar becomes the chief tree associated with spruce and silver firs. All ages are represented, but most of the trees are mature or (especially in the higher and less accessible places) over-mature. The density is varying, good on whole; but numerous small blanks occur. The reproduction is fair; many small thickets of deodar with scattered seedlings of spruce and other species. The aggregate area fully stocked with deodar is about 200 acres.

The only treatment in the past has been protection from fire. The best trees in the most accessible places were felled some years ago.

The enumeration gave:—

Unsound over-mature deodar		.	.	.	.	208 trees.
Sound deodar	over 2' diameter	.	.	.	509 ,,
,, ,,	$1\frac{1}{2}$' to 2'	,,	.	.	.	693 ,,
,, ,,	1' to $1\frac{1}{2}$'	,,	.	.	.	372 ,,
,, ,,	below 1'	,,	.	.	.	2,608 ,,
,, spruce	over 2'	,,	.	.	.	309 ,,

Remarks.—The removal of the over-mature and exploitable trees is urgently required in the interests of the younger stages of growth. The re-stocking of the blanks may be left to nature.

Stock maps.—Detailed descriptions of forest crops are tedious to write, and, in the case of a large area, bulky when written. Moreover, it is doubtful if any one ever derives much benefit from even the best descriptions that can be compiled for the large areas dealt with. The working-plans officer, with the aid of the map and area statements, forms his working-circles and determines the method of treatment more from the picture which an inspection of the various crops has left in his mind than from any written description; while, as regards officers who, personally unacquainted with the areas to which the plans relate, have to scrutinise the descriptions, their minds are incapable of grasping all the details and of forming a mental picture from such a mass of writing as usually accompanies an Indian working-plan dealing, as it often does, with a large area. It is, therefore, a question whether it would not be better, in some cases, to abolish these written descriptions and to replace them—apart from the broad general description of the forest or of each type of forest as a whole—by *stock maps*.

To prepare such maps it will, in most cases, be sufficient to distinguish each class of forest by a flat wash of distinct colour and to indicate the age by tone, the darker the tone the older the crop. Or a system of ruled lines may be used, different species being represented by different colours, density, completeness and age, by the closeness and length of the lines, and seedlings by small dots.

In the *Revue des Eaux et Forêts* for the 10th June 1890, Monsieur Marcel Volmerange writes as follows on the subject :—

" One of the most troublesome portions of a working-plans report is the description of the compartments. This requires as much minute care on the part of the writer as it does steady attention on the part of the reader ; and, after all, it is extremely difficult, from a perusal of it, to gather any idea of the forest in its entirety.

A graphic representation of the nature and composition of the forest would probably give a better general idea of its condition and contents, and could be prepared with far less trouble.

Such a method would consist in indicating on a sketch map by conventional signs the principal factors of the crop, different colours being used to show the species of trees, for example :—

Seedling crop	o	o	o	o	o •
Thicket of saplings	8	8	8	8	8
Pole crop	+	+	+	+	+
Young high forest	×	×	×	×	×
Mature high forest	—	—	—	—	—
Over mature or decaying high forest	=	=	=	=	=

Complete crops might be indicated by a continuous line

These conventional signs, or such others as might be preferred, could be used in various combinations, and would thus enable the composition of the forest in each compartment to be shown with whatever degree of accuracy or detail might be desirable ; while a general idea of the forest as a whole could be gained from a simple inspection of a map.

In order to complete the description, a brief account would be sufficient to explain the nature of the soil, unless it was thought practicable, and not too difficult, also to record the quality of soil by a similar method of signs and colours.

It would also prove interesting, after the lapse of a certain number of years, to make, using similar signs, a new map, which, by comparison with the old map, would at once indicate the alterations and improvements in the condition of the crops and would facilitate the determination of the changes required in the treatment."

Colonel Wilmer, of the Survey of India Department, in surveying the forest of the Central Provinces instituted some years ago such a system as the above at the instance of Mr. Mackenzie, the Chief Commissioner. Colonel Wilmer described as follows his system, which has since been officially prescribed for all Indian forest surveys :—

" The classification of forests and soils was adopted and carried on at the same time as the original detail survey was made, the former by the *colours* of lines used for shading, and the latter by the *direction* of the lines. The classification was shown on tracing cloth by symbols as follows :—

The forests were divided into five classes, *viz.*:—

1. Forest where teak abounds without bamboo . . . . Carmine.
2. Ditto   ditto   ditto   with bamboo . . . . Blue.
3. Miscellaneous forests . . . . Brown.
4. Grass lands . . . . Green.
5. Cultivation . . . . Yellow.

Soils were divided into four classes, *viz.*:—

1. Very good and rich . . . . . *Lines drawn vertically.*
2. Medium . . . . Do.   horizontally.
3. Very dry . . . . . . Do.   diagonally.
                                    from N.-W.
                                    to S.-E.
4. Unproductive . . . . . . Do.   do. from N.-E
                                    to S.-W.

Each Sub-Surveyor kept an outline trace of his daily work on a piece of tracing cloth, and also made notes of the nature of the soil and class of forest that he met with. At the end of a few days, when sufficient detail survey had been completed, he coloured up the portions and marked them with lines according to the fixed symbols. This trace the Sub-Surveyor kept going until his work was done, and it was examined at the same time as the topographical detail. Some differences of opinion were found at the adjoining edges as regards nature of soil, more especially between the 'medium' and 'very dry;' but these were reconciled.

On compiling the different Plain Table Sections into Standard Sheets, it was found that laying colour on to the tracing cloth spoilt it and rendered it opaque. The plan of drawing in the lines in their proper colour and direction was then adopted, and was found to answer all purposes, whilst the tracing cloth was kept clear and smooth."

In recording the nature of forest growth and soils the map should be on a sufficiently large scale. For such work great accuracy of topographical detail is not required, and therefore existing maps or enlargements of them will usually serve the purpose.

VALUATION SURVEYS.

When required : methods in use.—In many cases, when for instance the possibility is expressed by area, a good description of the crops is all that is requisite. But when the possibility is expressed by number of trees or by volume, as in the regular and selection methods, an enumeration or estimate of the number of trees is necessary. In the method of successive fellings the enumeration must also be accompanied by an estimate of the cubic contents of each tree, in order to prescribe the principal fellings. In such cases the analysis of the crop is not complete until the enumeration has been carried out.

The material in the standing crop may be ascertained accurately by *complete survey*, that is to say, by counting and measuring the individual trees over the whole area. Or it may be estimated, more or less accurately as desired, by counting the stock on *sample plots* of known extent or on *lines* of known width run through the crop; after which, by means of simple proportion, the number of trees in the whole area can be deduced.

Choice of a method.—The circumstances of each case must decide which of these methods should be adopted. Obviously the most correct is to count and measure each tree in the crop. But this involves the expenditure of so much time and labour that, except where a very accurate plan is required, and is justified by the high revenue expected from the forest, it is generally best not to employ it. It may, however, suitably be used when, as frequently happens, only one or two species irregularly distributed in a mixed crop are saleable, and when the plan is to provide for the working of those species only. In such cases the cost is not unduly high, and the results obtained from countings over sample areas are unsatisfactory.

The chief considerations which should determine the method to be adopted may be summed up as follows :—

The purpose of the survey and the degree of accuracy demanded in the plan.—It may be only necessary to ascertain the average production per acre, so as to calculate the yield in the manner described in the selection method. It is usually unprofitable to attempt to calculate the material on the ground with extreme accuracy; and, in any case, even complete enumeration and cubing of each individual tree might not ensure accuracy.

The nature of the crop.—Enumeration by sample plots should be avoided in the case of very irregular and open crops, or in crops containing only scattered trees for enumeration as in the case of selection-worked areas in which only one or two species sparsely distributed in a mixed crop are exploitable. Again, the method is unsuited to the case of very small areas; as there would be difficulty in selecting plots representative of the whole.

Method of sample plots; selection of the plots.—Where the conditions are such that the method of survey by sample plots is justifiable, separate plots should be marked off in each distinct type of growth. They should be chosen so as to represent fairly the average conditions of the crop or particular type of growth of which a pattern is taken.

Linear Survey.—In some cases it may be more convenient to use sample lines taken at regular intervals through the forest. Where the crops are mixed and variable, the position of the lines should be traced on the map on which the limits of the different types of forest are shown, and the results should be calculated separately for each crop through which the lines run. The width of the lines should be fixed so as to be convenient for purposes of calculation.

Thus a line 1 chain broad would represent 6·1 acre per chain-length; and, if the type of forest through which the line led was found to measure, say, 6 acres, while the length of line was, say, 9 chains, the number of trees could at once be found as follows:—

$$\frac{\text{number of trees in crop of 6 acres}}{\text{number counted on 9 chains of line}} = \frac{6 \text{ acres}}{0\cdot 9 \text{ acres}}.$$

The following rules * should also be observed in the case of sample plots. Their application to linear surveys need not be discussed.

(1) No sample plot should be selected on the edge of a crop; for a true average will seldom be found there.

(2) On slopes presenting a wide range of elevation, or in crops offering a great variety of aspects and soils, several sample plots judiciously distributed should be selected.

(3) The form of the plots should be regular: in general the best shape is that of a long rectangle.

(4) The aggregate area of the sample plots should not be less than 5 per cent. of the total area of the crop or type of growth of which they are chosen as the pattern.

(5) In mature crops no sample plot should usually be less than from 3 to 5 acres in extent; and in no case should the area be less than one acre, except, perhaps, in young uniform crops.

(6) In crops of considerable extent several plots should be taken and not one of large size.

Size-classes of the trees enumerated.—Whether the trees are counted over the whole area or only on portions of it, the number to be measured is so great that it is not practicable to record the correct diameter of each individual stem. Trees of more or less the same sizes are, therefore, grouped together in girth or diameter classes.

Where, as often happens, timber is sold by, and tables are prepared for, girth measurements, the classification might with advantage be by girths; the calipers used being marked so as to indicate girths not diameters.

The range of girth or diameter included in each class

---

* NOTE—These rules, and much of what follows in this chapter, have been extracted from the "Treatise on the mensuration of timber and timber crops," compiled from the German by Mr. P. J. Carter, and obtainable at the Office of the Superintendent, Government Printing, India.

should vary with the degree of accuracy sought and the size attained by the trees themselves. In very accurate surveys, differences of diameters as low as from 1 to 2 inches for large trees and from ½ an inch to 1 inch for small trees is sometimes taken; but such minute work is out of place in India, where a variation of 6 inches in diameter can usually be employed with advantage. Theoretically, the height ought also to be taken into account in making this classification; as height even more than diameter is influenced by the local factors of production. But this is impracticable in most cases, and diameter-classes alone are sufficient. Where, however, heights should be recorded, they can be estimated by eye or with the aid of instruments specially constructed for the purpose. Hitherto in India the classification usually adopted has been the grouping together of all trees whose diameters do not differ by more than 6 inches. Thus trees having a diameter of 6 inches (or a girth of 1½ feet) and under from one class; those of from 6 inches to 12 inches in diameter, or from 1½ to 3 feet in girth, another; those of from 12 to 18 inches in diameter, or from 3 feet to 4½ feet in girth, a third; and so on.

It would, perhaps, be more logical to classify according to the exploitable size of each principal species. In most mature forests of the sort generally met with there are three natural classes of stems more or less easily recognisable, viz., (1) exploitable trees of and above the exploitable dimensions; (2) medium-aged trees ranging from the minimum exploitable diameter down to, say, two-thirds of that size; and (3) young and suppressed stems, the diameters of which are less than two-thirds of the exploitable size. It is possible that the adoption of some such uniform system would be advantageous.

The enumeration of the trees.—The enumeration and measurement of the trees is usually performed by parties consisting each of one recorder and from 4 to 8 measurers, the latter being provided with calipers for measuring the diameters of the stems. The form of calipers in common use consists of a graduated rule (A A), at one end of which is a fixed arm (B B), and on and around which slides another arm (C C). To enable this arm to move freely, the hole (a, b, c, d) is made oblique but in such a manner that, as soon as the arm (C C) comes in contact with the log or tree to be measured, the arm assumes a position at right angles to the scale which it touches at the edges (c) and (b).

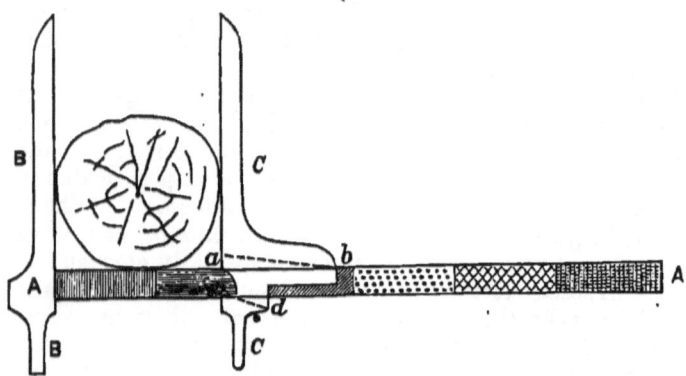

Instead of being graduated in the ordinary way, the rule (A A), in calipers intended to be used by illiterate workmen, may, with advantage, be painted in different colours at intervals equal to the differences between diameter or girth classes, so that each colour corresponds to a class. It has also been found a wise precaution to attach to the trees measured paper labels of the same colours.

Thus, supposing there were four classes of trees, the colours might be—
  Class I,   diameters above 2 feet, corresponding colour, white.
    "   II,   "   1½ to 2   "   "   "   green.
    "   III,   "   1 to 1½   "   "   "   red.
    "   IV,   "   ½ to 1   "   "   "   blue.

Where trees of different species are separately recorded, the shapes as well as the colour of the labels may vary. Thus species A may be marked with square tickets, species B with triangular tickets, species C with round tickets; and so on.

Each enumerating party consists of one recorder and of as many gaugemen and coolies as are necessary. The recorder is provided with a pencil and with a blank book ruled in columns for the required number of diameter classes. The columns may be sub-divided so as to show separately sound and unsound trees and kinds of trees to be separately recorded. The vertical columns are headed by the names of the colours on the beam of the gauge which correspond to the diameter classes, and each gaugeman is provided with a bag, containing in separate pockets the tickets of different kinds and shapes, and with a gauge or calipers. The gaugemen advance in line and measure each tree at breast-height from the ground. After a tree is measured the gaugeman calls out its native name, the colour denoting its diameter class, and, in the case of the highest classes, whether it is sound or unsound. The recorder then makes the necessary entry in his book, noting each group of ten trees separately. This record can suitably be made by two vertical rows of four dots, each row crossed diagonally by lines from corner to corner, as in the example on page 50.

The soundness or otherwise of a tree cannot always be determined, and many trees which are unsound at the base may be sound throughout the greater part of the stem. Trees are generally classed as unsound if they are noticeably decayed or hollow in the stem, if they have half-dead crowns, or are stag-headed; or if they give back a hollow sound when struck with an axe by a man standing on the ground.

Where the enumeration is to be complete, and not by linear survey or sample plots, the record should be prepared separately for compact areas not exceeding a few hundred acres in extent. For the purpose of checking the enumeration, certain areas or plots should be selected for re-survey, and the officer in charge should be present during at least one day's recording in each plot and should add the record made by himself to that made before his arrival. By keeping a careful account of the number of tickets issued and of the number returned, the balance that has been expended of each kind is known, and these numbers afford an additional rough check on the number of each kind of tree. But, unless the countings are actually verified, tickets may be either accidentally lost or intentionally made away with, and the record may be completed accordingly so as to lead to the belief that a large amount of work had been accomplished. Constant and careful supervision by a trustworthy person is, in fact, the only means by which reliable results can be secured.

It is usually the custom to blaze with an axe the trees measured. This, however, for obvious reasons, should not be done so as to injure the stems in the manner usual with native workmen. Where water is available, the trees may be marked with whitewash, *made with rice water*, so as to render it adherent; but ordinarily a light blaze on the bark of each tree will suffice.

In crops consisting of fairly regularly shaped and not very large trees the measurement of a single diameter may suffice, especially if the crop is traversed by the enumerating party in various directions. But otherwise the mean of two diameters, taken more or less at right-angles to one another, should be recorded. Although a matter of petty detail, it is necessary to observe that the calipers should be properly applied to the trunk of the tree and the diameter read off before the instrument is removed. The diameters should be all measured at breast-height, and on hill-sides this height should be taken on the upper side of the stem. Breast-height has been assumed to be $4\frac{1}{2}$ feet; but as the boles of trees do not taper either regularly or very rapidly, it is not necessary that this height should be exactly measured before the calipers are applied. Sufficient accuracy is attained if the measurer is careful to hold the calipers horizontally at the height of his chest, and if the diameter is measured at any height between 4 and 5 feet, and at a place where the stem is free from excrescences and branches. When a tree divides into two or three main stems, near the point at which the calipers would ordinarily be applied, each stem should be measured separately.

Where the enumeration is to be complete, the survey should, if possible, be effected over successive narrow strips, each strip being gone over once and in a direction opposite to that in which the immediately preceding strip has been surveyed. On steep slopes it is convenient to run the strips horizontally and to begin at the bottom of the slope. The number of measurers that can keep one recorder fully employed depends of the density on the forest, on the nature of the ground, and on whether all or only certain classes of the trees composing the crop are to be measured. The number of measurers for each recorder may then, according to circumstances, range from 2 to 6, or even 7 or 8. As the survey progresses, the trees measured are immediately marked. The mark should be made on the side towards the area still remaining to be surveyed, so that, when the next strip is being examined, the men can at once recognise up to what point the strip just completed extends. The duty of the recorder is to see that all the trees are measured, that the calipers are properly applied, the diameters read before the calipers are removed, and the mark made on the correct side of each tree measured. He should also note the distinction drawn by the measurers between sound and unsound trees, and he should keep his men, as far as may be, in sight and in line. When a division into height classes is necessary, he has also to measure, or to estimate by eye, the height class of each tree as it is gauged.

Recording the results of the enumeration.—The following is a sample of a convenient form of a field-book for enumeration surveys, in which the classes include a range of diameters and the forest is very irregular. Some of its advantages are :—(i) it requires very little ruling, (ii) it may be easily prepared from day to day by the recorder himself, and (iii) as the width of its different columns can, for that reason, be varied to suit the composition of the crop to be surveyed, a whole day's work, comprising the record of several thousand trees, can be compressed into a single opening of the book :—

Forest.	Block.				Compartment.	
	Diameters.				Total number of trees.	Remarks.
Species	6 inches to 12 inches.	12 inches to 18 inches.	18 inches to 24 inches.	24 inches and over.		
Sál	⋈⋈⁖	⋈⋈⫶	⋈	⋮	68	
Sain	⋈⋅	⋮⋅	⋮		19	
Miscellaneous	⋈	⋈⋮	⁖		30	

The numbers of trees are shown by dots and strokes. Each group, representing ten trees, consists of two upright rows of four dots each, joined by two diagonal lines which represent, respectively, the ninth and tenth trees Thus, ⋈ =10, ⫰ =9, ⁛ =8, ⁚⁚ =7; ⁙ =6; and so on.

CALCULATION OF THE VOLUME OF MATERIAL.

Type-trees and form-factors.—The number of trees in each size-class having been ascertained, either by counting and measuring each individual tree in the whole area or by means of sample plots or lines, the volume of material in each class and, consequently, in the whole crop, may be calculated when required by means of *type-trees*, as follows :—

We may assume that, on an average, in one and the same crop, trees of the same diameter and height have the same cubical contents. For all practical purposes, therefore, we may assume, without any great error, that if we select trees that are *representative of each size-class*, ascertain their contents and multiply this figure by the number of trees in the class each represents, the sum gives the volume of material in the whole crop. Such trees are called *type-trees*, and may be selected either by eye, by taking an *average-tree*, fairly representative of the crop or class as regards size, or, more accurately, as follows :—

Calculate, with the aid of tables, the sectional areas at the base, where measured, of all the trees in the size-class and total the whole. Divide this figure by the total number of trees in the size-class, and the quotient will be the basal area of the type-tree required. Calculate the diameter corresponding to this basal area; fell several trees of this diameter and ascertain their contents by actual measurement. The mean of their aggregate contents multiplied by the total number of trees in the size-class will give the total volume of material in that size-class. Proceeding in the same way with each size-class, the sum of the volumes obtained gives the total volume of material in the whole crop.

By *basal area* is meant the area of the section of the stem of a tree measured on a plane at right-angles to the axis of the stem and at breast-height, or, as is usually taken at 4½ feet from the ground.

This area is approximately equal to one-fourth the square of the mean of the longest and shortest diameters of the tree at this height multiplied by 3·1416.[*] For the mean diameter is $\left(\frac{D+d}{2}\right)$, where D and d are respectively the greatest and least diameters of the stem.

By *form-factor* is meant the proportion which the true contents or volume of a tree bears to a cylinder having the same basal area and height. The volume obtained by multiplying the basal area by the height of a tree is of course that of a cylinder; but, as trees taper, the true contents of the stem are something less than this. If $a=$ basal area of the trunk at breast-height, $h=$ the height of the bole or length of stem taken, then the volume (C) of an ideal cylinder whose basal area is a and height

---

[*] Value of π

h, is $a\ h$, and we have the following formula with regard to the constant or reducing fraction called the *form-factor* (f) and the true volume (c') of the stem :—

$$c' = f(C) \qquad (C) = a\ h\ ;\ c' = f\ a\ h\ \therefore\ f = \frac{c'}{a h} = \frac{c'}{(C)}$$

The value of f is determined by felling a sufficient number of trees and by ascertaining practically the value of $\frac{c'}{(C)}$, that is to say, the relation which subsists between the ideal cylinder and the true volume of the stem. With the above explanation we may proceed to prove the rules with regard to the selection of type-trees as follows :—

> Let C = the total volume of all the trees in a size-class.
> A = the sum of the basal areas of all the stems.
> H = the average height of the tree.
> F = the average form-factor of all the trees.
> n = the number of trees in the size-class.
> c' = the contents of the type-tree.
> a = the basal area of the type-tree.
> h = the height of the type-tree.
> f = the form-factor of the type-tree.

We require to find a, the basal area of the type-tree representing the size class.

By supposition $C = A\ H\ F$, and $c' = a\ b\ f$: also $c' = \frac{C}{n}$.

It follows that $c' = \frac{A H F}{n}$ and, therefore, that $a\ b\ f = \frac{A H F}{n}$. Therefore, assuming that $h\ f = H\ F$, $a = \frac{A}{n}$, or, expressed in words, that *the sum of the basal areas of all the trees in the class divided by the number of trees in that class equals the basal area of the type-tree.*

As the basal area $= \frac{\pi}{4} \left(\frac{D+d}{2}\right)^2$, it follows that the mean diameter of the tree sought $\left(\frac{D+d}{2}\right)$ is equal to the square root of the basal area of the type-tree, as calculated above, divided by one-fourth the value of π, or the mean diameter $= \sqrt{\frac{4 a}{\pi}}$. This, in practice, would be found from a table.

THE MEASUREMENT OF TREES AND LOGS.

General rules.—The following directions in regard to measuring trees are taken (with some considerable omissions) from the treatise on the measurement of timber and timber crops already referred to.

The *heights* of standing trees are ascertained by means of special instruments, designed for that purpose and known as dendrometers and hypsometers. These instruments are of two kinds—

(i) those which give the height without calculation, their construction being based on the principle of similar triangles ; and

(ii) those which give the angles made with a horizontal line by the lines of sight to the top and base of the tree.

For measuring *lengths* graduated rules or tapes may be used. Where great accuracy is required, the length of a felled tree or log should be measured paralled to its axis and not on its sloping surface. The sectional area of a log or tree can very rarely indeed be obtained directly. In nearly every case the girth or diameter must be measured, and the area of the section determined as if the section were a circle. Area of section $= \frac{3.1416}{4}$ (diameter)².

Girths are measured with tapes. It is convenient to employ tapes graduated on both sides, one side for reading the girth and the other for reading the corresponding diameter. The zero end of the tape should be furnished with a sharp metal point, which can be easily fixed in the bark of the tree, so that one person may be able to measure any stem, no matter how thick. As a circle encloses a greater area than any other plane figure of equal perimeter, and as the sectional outline of trees is seldom quite circular, the contents of a log or tree calculated directly from the girth will usually be in excess of the true contents. Unless the contour of the log is circular, it is impossible to obtain by girth measurement the circumference of the circle which encloses the same space as the section whose area is required. Irregularities of outline, due to fluting, bark, etc., cannot be overcome in measurements of girth, whereas they can more or less successfully be allowed for in measuring diameters. Experiments made in Baden prove that girth measurement yields a result from six to ten per cent. greater than that obtained by means of diameter measurement It is, however, obvious that, in cubing logs which depart from the cylindrical form, the measurement of the girth is more to be relied on than the measurement of a single diameter. When the contents of a log are to be deduced from diameter measurement, that diameter should be sought which, considered as the diameter of a circle, gives a result, as nearly as practicable, equal to the area of the section measured. When the section is elliptiform, the mean of the longest and shortest diameters should be taken.

The mean of these diameters D and d is $\frac{D+d}{4}$ and the area of the section is then assumed to be $\frac{\pi}{4} \times \frac{D^2 + 2Dd + d^2}{4}$. Now, as the real area of the ellipse is $\frac{\pi}{4}$Dd, the mode of measurement recommended gives an excess of $\frac{\pi}{4}\left(\frac{D-d}{4}\right)^2$; that is to say an excess equal to the area of a circle whose diameter is equal to half the difference of the two measured diameters. Save in very exceptional cases this difference is small enough to be negligible. The area of sections of irregular contour can be determined from the mean of three diameters; but the result thus obtained will generally be found to be somewhat too high.

Diameters are measured with calipers such as the instrument already described at page 47. This instrument, it will have been noticed, resembles in all its essential parts a shoemaker's measure. In measuring logs and trees the following general rules should be borne in mind :—

    (i) Diameters, in experiments and in accurate valuations, are to be preferred to girths.
    (ii) In the case of stems of elliptical or oval shape, take the mean of the largest and smallest diameters.
    (iii) In the case of all large stems measure at least two diameters.
    (iv) In the case of stems of irregular shape measure several diameters and avoid all protuberances, etc.
    (v) Measure diameters and girths always in a plane at right angles to the axis of stem.
    (vi) If the place of measurement falls on an irregular part of the stem, measure the diameter or girth, as the case may be, at equal distances above and below the irregularity and take the mean of the two measurements.
    (vii) Moss, etc., thick enough to vitiate the measurement of the stem, should be removed.
    (viii) If a very accurate measurement of an irregular section is required, transfer its outline to tracing paper and compute its area with a planimeter or acre-comb.
    (ix) Felled trees should be divided in the usual way, that is to say into logs and smaller pieces, and should then be measured.
    (x) Never be without tables showing at a glance the areas of circles for given diameters and girths.

Several formulæ, approaching more or less to accuracy, have been devised for the determination of the contents of *round timber*; but only two are of practical utility. They are :—

    (1) volume of contents of log = half the sum of the areas of the top and bottom sections × the length;
    (2) contents = area of mid-section × length.

Both formulæ contain an error, the extent of which is proportionate to the amount of difference between the area at the top and base, respectively, of the log, that is to say, to its degree of taper; and this error increases as the square of that difference. The second formula always gives too small a result, the first too great, the error of defect in the one case being one-half the error of excess in the other. The second formula has also another advantage for which it is to be preferred. The ordinary modes of measurement and calculation give, as a rule, too high a figure for the sectional area concerned in each case. This excess is partly compensated by the employment of the second formula; whereas the use of the first would only exaggerate it. In order still further to diminish error long logs should be measured in two or more sections, the number of the sections increasing; i.e., their length diminishing, with the taper of each log. The contents of logs of regular shape and not exceeding 20 feet in length may, however, be deduced from their sectional area in the middle. Longer logs, even if of regular shape, should be cubed in two or three sections. All large round logs should be measured singly. If the logs are stacked so that they cannot be conveniently measured in the middle, the mean of the sectional areas at the base and at the top must be taken. The mean sectional area should not, under any circumstances, be deduced from the mean of the diameters at the two extremities, respectively, or an error of from 10 to 15 per cent. may result. Poles are seldom cubed singly; but nearly always in stacks, built up of poles of one and the same length and of approximately the same diameter. Their solid contents are generally ascertained by inspection from special tables. Straight and regularly-shaped branches are measured in the same way as logs.

It must be observed, however, that while diameter measurements give more nearly the actual contents of logs and trees, yet in practice the contents of round timber are always calculated, for trade purposes in India and in England, on the assumption that the sectional area is arrived at by squaring the quarter girth.

Square cut timber must, of course, be cubed by the formula: volume = length × width × thickness.

The solid contents of *small pieces*, toppings and loppings and irregular-shaped pieces from stumps and roots, are obtainable by the *water-method* (their volume being equal to the quantity of water they displace when submerged), or by the water-method and weighment combined. For the water-method special vessels, called *xylometers*, may be employed. In the combined system, samples of each kind or class of wood are successively weighed and measured by the water-method, and the contents of the entire quantity in each class are then worked out by means of simple proportion. Figures expressing specific gravity cannot be employed; since the specific gravity of wood varies—not only according to the amount of moisture present but—even in one and the same tree according to the part from which the wood has been derived.

The most rapid way of measuring small wood on a large scale is to stack it cut up into billets of the same length, the width of each stack being equal to the length of the billets. The contents of a stack will be equal to length × height × common length of the billets. The length of a stack built up on a slope must be measured horizontally. The above formula will give us only stacked contents. To reduce the latter to solid contents, we must determine, by the water method or by the combined water and weighment method, the exact volume of a sufficiently large number of stacked units, thereby obtaining the ratio between solid contents and stacked contents. To obtain the solid contents of a stack it is then necessary merely to multiply the stacked contents by this ratio which may be termed the *reducing factor*.*

In connection with the determination of the solid contents of stacked wood it is obvious:—

(a) That the longer the billets or the less carefully built up the stacks, the less will be the solid contents. In careless stacking, billets often lie across one another.

(b) That the thicker or more regular-shaped the billets, or the more carefully built up the stacks, the greater will be the solid contents.

(c) That the larger the stacks, the greater will be the reducting factor.

When *bark* is sold separately its quantity may be

---

* The following figures may be accepted as average factors for converting stacked into solid contents—

Split wood	0·65 to 0·80
Round billets (large)	0·50 to 0·65
Ditto (medium)	0·30 to 0·50
Small stuff, stumps and roots	0·15 to 0·30

determined either by weighment or by ascertaining the volume. The solid contents are calculated by reducing factors in the same way as the solid contents of small wood. Experiments give from 0·3 to 0·4 as the average factor for bark. It has been found that the quantity of bark varies from 6 to 15 per cent. of the total volume of the tree or crop.

There is in India only one practicable method of arriving at the contents of *standing trees*, namely by means of form-factors. It has already been explained that the cubical content of a tree is equal to some fraction (f) of the ideal cylinder whose basal area is equal to the section of the tree at breast-height and whose height is that of the stem.

Form-factors may be deduced, according to the requirements of the case, for the whole tree, for the stem only, or for the branches.

It has been supposed that the sectional measurements have been taken at the height of a man's chest, assumed, for the sake of uniformity, to be 4 feet 6 inches above the ground. But it is obvious that any other conventional height would serve the purpose, although it is usual and most convenient to employ the one we have adopted. We need refer to only one other convention which is sometimes used. The diameter may be measured at a constant fraction (say, for instance, one-twentieth) of the height of the tree, in which case the form-fractors obtained are termed *normal*. Normal form-factors yield perfectly correct results; but they are not practical owing to the difficulty and trouble of measuring at various heights from the ground. Form-fractors are said to be *absolute* when the base of the ideal cylinder is assumed to be in the same horizontal plane as the diameter or girth measured. In this case the contents of the portion of the stem below the plane must be calculated separately.

As may be supposed, calculations based on form-factors give better results for an entire forest than for individual trees. The preparation of a complete set of form-factors requires great care and experience, as ultimate accuracy depends entirely on the selection of the type-trees, whose dimensions serve as the basis of all the calculations. In some cases the trees of a crop have been grouped into various classes according to their height and shape, and a separate form-factor calculated for each class. The most recent investigations prove that form-factors vary chiefly with the height of the trees.

CALCULATION OF THE EXPLOITABLE AGE.

General rules.—In order to calculate the exploitable age of a tree it is necessary to ascertain the girth or diameter at which it furnishes the greatest quantity of the most useful material that is required. As a rule the price is the best gauge of the utility; so that in most cases it may be said that trees are exploitable when, after deducting all expenses of exploitation, the price they realize per unit of volume is the highest obtainable.

It not infrequently happens in India that it has to be decided—not what is the price, or what, is the most useful produce but—whether the people in the vicinity of the forest or the general community should enjoy the produce. The people in the neighbourhood may require small wood-fuel and pasture for their cattle; the general community, large timber. As a rule, the decision goes in favour of the local wants, though this could not always be defended from a purely economic point of view. It is, moreover, frequently overlooked that in growing large timber small timber from the thinnings and branches is also available generally in as great abundance as can be consumed within the radius to which it may be profitably transported; when more grazing can be provided than if the crop were cut when young and the forest, consequently, can be closed more frequently to cattle.

By *price* is here meant, of course, the *net* price of the trees when standing in the forest, after deduction of all cost of felling and extraction. The price must be calculated per unit of volume in the rough, unless where there is only a demand for standing trees. When poles or timber in the rough or logs are sold, the price realised per cubic foot for differently sized pieces directly indicates the size of the trees which are most useful and which will bring in the highest revenue; provided that the cost of extraction is in all cases previously deducted.

It will often be found that, owing to defective means of transport and to the greater cost consequently involved in extraction, the *net* price of large logs is lower than that of small, although the selling price of the former when delivered to the consumer may be very much higher.

When the size of the trees is calculated from the ruling price for converted timber, planks, sleepers, etc., or manufactured articles, the loss in conversion must be calculated in order to ascertain the cost per cubic foot, and that size should be ascertained in respect of which the loss is least. Almost invariably the loss will be least when the trees are largest.

Suppose that sleepers, each containing 2 cubic feet and costing one rupee to saw and deliver, sell for R3 each. In order to ascertain the price realised per cubic foot in the rough, it will be necessary to know the average number of sleepers yielded by and the average cubic contents of an exploitable tree. Suppose the average diameter of the trees felled was $1\frac{1}{2}$ feet and that it was found that each tree felled contained 80 cubic feet and yeilded 20 sleepers, the price realised per tree would be $20 \times 2 =$ R40, and the price per cubic foot standing would be R$\frac{40}{80}$ = 8 annas. The loss in the conversion of these trees is 80—40 or 50 per cent. It should be ascertained whether this loss would not be less and the price realised consequently higher if larger trees were felled. Thus, suppose it was found that 2 feet trees containing 120 cubic feet yeilded 40 sleepers, the net price realised per standing trees would be $40 \times 2 =$ R80, or per cubic foot R$\frac{80}{120}$ = 11 annas. The loss on conversion is 120—80 = 40 cubic feet, or 33 per cent., as compared with 50 per cent. when $1\frac{1}{2}$ feet trees were felled.

A higher revenue is not realised by felling large trees unless the net price of the latter per cubic foot is higher. The price per tree standing would of course be higher even if the price per cubic foot did not rise. But this must not be mistaken for a higher revenue; for on the same area more small trees than large can be grown. The larger trees might bring in double as much per tree as the smaller; but there might be double as many of the smaller stems. The quantity of material produced per annum would be the same.

In fixing the dimensions of the exploitable tree the number of stems to be felled is decided upon; and it might be thought that, because the lower the age of felling the more trees can be felled, the adoption of a lower age would be better. Thus, suppose a forest of 1,600 acres in which the exploitable size is fixed at 18 inches in diameter corresponding to an age of 100 years; that the annual yield is, say, 1,184 trees; and that it is proposed to raise the exploitable size to 2 feet diameter. We will assume that the average production of the soil per acre is 63 cubic feet a year, and that the tree of 2 feet contains 135 cubic feet, as compared with 85 cubic feet in the $1\frac{1}{2}$ feet tree. The number of trees that could be felled annually would be $\frac{85}{135} \times 1,600 = 747$. It might be argued that it would be better to fell 1,184 trees a year than only 747. This depends on the purpose for which the trees are required and on the price realised. Thus, if, owing to the better wood or less loss in conversion, the net price realised per cubic foot for trees of the larger size were 4 annas, as compared with 3 annas for the smaller trees, each of the larger trees would be worth R34, as compared with R16 for the smaller; and the annual revenues would be R25,398 or R18,944.

Indian forestry is not ripe for elaborate calculations, and must be satisfied with felling when the revenue will be highest or the produce the most useful: otherwise it would also be necessary to consider the greater capital involved in producing the larger size timber in view to taking account of the rate of interest on that capital.

The correct calculation of the exploitable size is of the greatest importance and demands a good deal of careful local enquiry and comparison as to selling rates, etc. In European countries, where forestry in all its branches has long been practised, and where the wood trade is fully developed and established, the most profitable size is well known for each class of produce and each kind of treatment; but this is not the case in India.

The following is an example of the sort of calculation that would be made to ascertain the most profitable size for felling for fuel:—

Description of material and its thickness.	Selling price per 100 cubic feet stacked.	Cost of cutting and extracting per 100 cubic feet stacked.	Net price per 100 cubic feet stacked.	Reducing factor per 100 cubic feet stacked.	Net price per 100 cubic feet solid.	REMARKS.
	R	R	R		R	
Fuel in billets of over 10″.	5·89	1·6	4·29	0·63	6·81	
Fuel in billets of 10″ to 6″.	5·32	1·0	4·32	0·57	7·58	
Fuel in billets below 6″.	3·50	1·51	1·99	0·37	5·38	

It is evident from this that, assuming the average annual increment to remain fairly constant, there is every financial advantage in growing trees that will furnish billets of from 10″ to 6″ in diameter.

The size having been decided it only remains to determine the age at which the trees attain that size. This is often exceedingly difficult in India, owing to the ring marking being indistinct, and also to the fact that several species form more than one concentric ring of wood each year. In such cases the rate of growth can only be determined when trees of known age are to be found. The following remarks apply, therefore, only to those species which form a single distinct concentric ring of wood each year.

In the case of *felled trees*, the required age is found by counting the number of concentric rings on the section of the stool or base of the trunk. To ascertain the diametral increase of growth, the countings should be made along several radii and the mean should be taken. Where necessary the rings in the sap-wood and bark should be separately counted and recorded, and all great extremes should be rejected. Usually, with a view to determine, as is often necessary in order to calculate the possibility in selection-worked forests, the number of years required by trees to pass from one size-class to another, the average number of rings included in the diameter of each size of class is separately recorded.

For a good many species the age may be very roughly determined without felling the trees by the use of the little instrument known as *Pressler's borer*, a hollow gimlet which, on being screwed into the trunk and then withdrawn, extracts a small cylinder of wood on which the ring-markings can be counted. From two or three such borings, on different sides of the trunk, the average number of rings per unit of radius can be ascertained.

This instrument is a gimlet, consisting of a tube (G) with a very sharp-cutting edge (E).

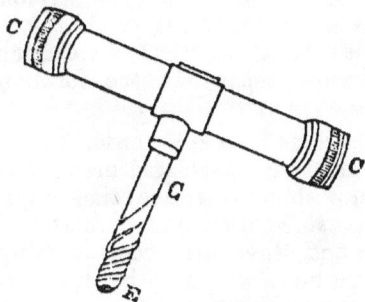

To render the instrument easily portable, the gimlet G can be taken off and placed within the hollow cylinder CC, of which the caps are removable.

As the tube is forced into the trunk of a tree a cylinder of wood is cut out. On withdrawing the gimlet the cylinder can be easily pushed out of the tube and the ring-markings on it counted. Well-formed fairly cylindrical stems of various girths should be operated upon, and care should be taken that the instrument proceeds into the trunk towards the centre.

CHAPTER III.—THE WORKING-PLAN.

PRELIMINARY EXPLANATIONS.

General arrangement followed.—The prescription of the possibility—the manner and locality in which it is to be exploited—constitutes, practically speaking, the *working-plan* for all forests of which the exploitable material is wood; and, as the manner in which the possibility is calculated depends on the method of treatment, it will be convenient to discuss separately the form of working-plan required under each method.

This will be done in regular order, the coppice methods, as the simplest, being considered first. The manner of prescribing the provisional treatment that crops may and usually do require, transformation and restoration fellings, preparatory thinnings and other improvement fellings, in order to apply these methods, will also be dealt with in describing the nature of the working-plan for each:—

Method of simple coppice.
,, ,, coppice-selection.
,, ,, branch-coppice.
,, ,, coppice with standards.
,, ,, clearances.
,, ,, "storeyed" forest.
,, ,, selection.
,, ,, successive fellings.
,, ,, pastoral treatment.

Before dealing with these methods in detail some preliminary explanations are, however, necessary.

Manner in which the possibility is prescribed.—It has already been explained that the possibility may be prescribed in three ways, viz., *by area, by number of trees* or *by volume of material*, and that either of the two latter may be combined with the former. Practically speaking this is what is always done, and the possibility is expressed by prescribing the felling, under certain sylvicultural rules, of the crop, or of a limited number of trees or volume of material *in a given area*.

General and special plans.—The prescription of the possibility involves the formulating of *a general plan* or framework on which the whole working of the forest is built up and a *special plan* or statement of the fellings to be made during a limited period. It has in some cases been imagined that the general plan might be dispensed with and the fellings prescribed in what has been called a "preliminary working-plan." This is, however, a mistaken notion. A forest exploitation must have a definite purpose which cannot be arrived at without considering what crop and capital is to be created. This necessitates the determination of the exploitable age and the drafting of a general scheme of working for that age.

Thus, suppose the working-plan for 1,000 acres of sâl forest to be—treatment by the selection-method with the object of growing sâl trees of two feet in diameter and requiring 100 years to attain that size; the possibility being 1,000 trees a year. These prescriptions constitute the *general plan*. The regulation of the fellings for a certain time, say for 10 years (that period having been adopted as the felling rotation) and their allocation to definite areas or blocks, each about one-tenth of the total area, would form the *special plan* or *statement of fellings* for the first felling rotation. A better example is, however, afforded in the method of successive regeneration fellings in which the periodic blocks and the order in which they are to be regenerated are prescribed in the *general plan*, while the *special plan* prescribes in detail the fellings to be made in each block during one period only.

Provisional working scheme.—The possibility is based on the condition of the existing crop; and as this crop is practically always abnormal, being either insufficient or superabundant or irregularly arranged, the possibility also exhibits corresponding divergencies from the normal standard.

Thus, a coppice of 500 acres is to be exploited at 20 years. The normal annual felling would extend over 25 acres of coppice 20 years old. But, if there were no crops of, say, 10 to 15 years old, it would not be possible during the first rotation to follow the normal plan: either the area occupied by the crops aged from 16 to 20 years should furnish the fellings for the first 10 years, and then a larger area would be felled, or crops of a lower age than 20 years should be felled during some years.

It has by some been considered necessary to distinguish between the actual scheme applied and the ideal one; and the former has, in consequence, occasionally received the special name of *provisional plan*. This is, however, hardly necessary; because, in the great majority of cases, the only plan with which we have to deal is provisional, and there can, therefore, be no confusion.

Preparatory period.—The length of time required in order to constitute the forest according to the normal condition is sometimes called a *preparatory period*. It is necessary to adopt a preparatory period where the age-classes are defective or irregularly arranged, where it is desired to hasten the felling of over-mature material. or where immediate regeneration of the crops cannot be undertaken. Theoretically the length of the preparatory period should be equal to the exploitable age; but, in practice, a shorter period is generally adopted, and the attainment of the desired end is hastend by judicious improvement fellings, etc. The object is to obtain a crop normally constituted according to the method of treatment adopted and containing a regularly graduated series of age-classes. Such a forest capital can generally be secured in a shorter time than that represented by the exploitable age.

Thus, the irregularly constituted coppice of 500 acres, already taken as an example which it is wished to exploit at 20 years, could only be constituted, according to the normal type, by making during the next 20 years 20 successive fellings of 25 acres in extent each year. After the lapse of 20 years there would be on the ground twenty age-classes, each occupying the same area and differing in age by one year. In practice, however, such a treatment is not necessary, and it is merely sought to properly constitute the capital as soon as possible. In the case of coppice exploited at a low age, the provisional period is, however, generally equal in duration to the exploitable age.

Prescribing the fellings.—The general working scheme is merely an outline; because detailed operations could not, in most cases, be prescribed with any degree of certainty for so long a time in advance as the exploitable age or the duration of the preparatory period. It is, therefore, necessary to regulate in a special statement and for a shorter time the—

duration of operations;

area to be operated on in each year or period;

order to be followed in the fellings;

nature of the fellings;

material to be removed.

Period for which fellings are prescribed.—The interval during which the fellings should be prescribed depends on the general working scheme, as provisionally modified, and on the nature of the plan. The interval should not be too long, as the longer the more likely is it that unforeseen events may necessitate its modification; and this may prove inconvenient. As a general rule, the term should be long enough to enable the prescribed fellings to pass once over the whole area to be worked, or to complete any definite series of operations already undertaken. That is to say, the fellings would generally be prescribed for one whole felling rotation in the case of coppice and selection-worked forests, for one period in the method of successive fellings, etc. In Europe longer periods than 10 or 15 years are seldom taken; but in India, in the case of selection fellings, 25 years or more have sometimes been found expedient. An interval of time varying from 10 to 30 years would, therefore, generally be that within which the prescriptions would run.

Area operated on.—The area to be operated on each year may be determined by dividing the number of acres in the whole forest, less blanks, etc., by the number of years in the felling rotation, and then by apportioning to each block or natural sub-division, according to the fertility of the soil, a certain number of coupes each of that area.

The following example will explain this process. Suppose a forest of 3,870 * acres containing a crop consisting of three types, only one of which is at present productive owing to the others containing no saleable species; also that four blocks bounded by roads, rivers or other landmarks, have been formed and analysed as follows:—

Name of Block.	Exploitable sâl forest.	Unproductive mixed forest without sâl.*	Blank.*	Total.	Remarks.
	Acres	Acres.	Acres.	Acres.	
Paimar	690	180	80	950	⎫
Datwind	510	100	60	670	⎬ Very good soil.
Haigla	530	250	60	840	⎭
Baira	1,240	150	20	1,410	Poor and rocky soil.
Total	3,000	680	190	3,870	

We will assume that the felling rotation adopted is 15 years, and that it is consequently required to divide the total area into 15 portions of *equal forest productivity*. This may be done as follows:—

Total area of forest . . . . . . 3,870 acres.
Deduct blanks and unexploitable areas . . . 870 ,,
Total exploitable area . 3,000 ,,

The average area to be operated on annually is $\frac{3,000}{15} = 200$ acres. In order to lay out these areas each block may be divided into an integral number of coupes containing an area of exploitable forest greater or smaller than the average according to fertility of the soil. Thus, Paimar, Datwind and Haigla, containing very good fertile soil at the base of the hills, might each be divided into coupes containing on an average, less than 200 acres of sâl forest; whilst Baira, situated in rocky ground in the hills where the quality of the soil is poor, might be divided into five portions, each

---

* It need hardly be explained that small areas of unexploitable forest would be included where necessary in a working-circle.

containing about 250 acres of sâl forest, or a good deal more than the average When marked out on the ground the coupes would, we will suppose, be as follows:—

Serial No. of coupe	Block.	Area in Acres.			
		Sâl.	Unproductive.	Blank.	Total.
1	Paimar	180	60	Nil	240
2	Do.	175	20	Nil	195
3	Do.	175	Nil	20	195
4	Do.	160	100	60	320
5	Datwind	185	70	30	285
6	Do.	175	30	Nil	205
7	Do.	180	Nil	Nil	180
8	Haigla	175	50	60	285
9	Do.	175	100	Nil	275
10	Do.	180	100	Nil	280
11	Baira	240	Nil	Nil	240
12	Do.	230	10	20	260
etc.	etc.	etc.	etc.	etc.	etc.
Total		3,000	680	190	3,870

In selection working, where the number of trees to be felled is prescribed, the usual practice in India has been to study the results of the enumeration survey and to select areas capable of furnishing the number of trees required each year. Thus if the possibility were fixed at 2,000 trees, areas capable of furnishing 2,000 trees would be selected to constitute the annual coupes. This system is, however, open to objection in that the coupes are not permanent as they ought to be. They depend merely on the crop for the time being. At the next felling rotation totally different annual coupes might have to be formed. Where a sustained annual yield is not of special importance, it would, therefore, be better to form permanent annual coupes and then to prescribe the fellings to be made in them.

Balancing the production.—In coppice fellings, and other cases in which the possibility is prescribed by area only, it is sometimes sought to *balance the production* and to determine the area of the coupes by processes more exact and scientific than those described above. The production depends on what has been called the *quality of the locality*, that is to say, on the influence of the climate, aspect, soil, etc., and on the composition and density of the crop; and it has been attempted, by assigning numerical co-efficients to each of these factors, to arrive at the result sought by means of mathematical formulæ. But such calculations, based on uncertain *data*, are often misleading and occasionally lend merely an appearance of mathematical accuracy to estimates which can be more correctly made by the exercise of a little judgment and common sense. In the present state of our knowledge of forestry in India they are entirely out of place.

Locating the fellings.—For locating felling rules have been formulated which are to be found in most works on sylviculture. These rules, although theoretically occupying an important place in organised forest working, are not, however, always applicable, especially in the case of irregular forests worked by the selection method. They may be stated as follows:—

(1) The fellings should be adjacent and succeed one another in the order in which made, and should have the most regular form possible.
(2) They should be so located that the produce of an area in course of exploitation need not be carried through the young crops in the portions of the forest recently felled.
(3) They should proceed from the side least exposed in a direction contrary to that of the prevailing dangerous winds.
(4) On steep slopes the fellings should be commenced at the bottom.
(5) In hill forests the coupes should be long and narrow in form, and have their longest sides perpendicular to the direction of the dangerous winds.

Nature of the fellings.—The nature of the fellings to be made depends on the permanent or temporary method of treatment, and may be explained by a single term. A list of the various fellings has already been given in connection with the methods of treatment. But, in the irregular condition of the forests in this country and owing to the want of well-known methods of cultural treatment applicable to them, it is always well to indicate briefly the more important points connected with the application of the fellings. This may be done by a few remarks in the statement of fellings itself, or in a separate note; or the detailed descriptions of the forest may be referred to, and such suggestions as are likely to prove useful to the executive officers may be entered in the remarks column of the statement for each compartment or block.

Material to be removed.—The possibility, as calculated by one or other of the methods to be explained hereafter, should be prescribed for the length of time for which the detailed statement of fellings is drawn up, and it may possibly be subject to revision during that period.

We may now proceed to discuss the manner in which plans may be suitably framed according to one or another of the various methods of treatment enumerated at the beginning of this chapter.

THE METHOD OF SIMPLE COPPICE.

General plan.—The general plan in the case of the simple coppice method is exceedingly simple. It consists in dividing the forest into as many annual coupes as there are years in the exploitable age, and prescribing the felling of one such coupe in rotation each year or period. Where the age-classes are not suitably distributed or graduated, a provisional plan is necessary; and, as the age of exploitation is short, the length of the preparatory period during which the provisional plan remains in force is usually the same as the exploitable age.

Exploitable age.—The age of felling is the first point to determine. This age varies within tolerably narrow limits. The trees cannot be felled at a very advanced age, or they will have lost the power of throwing out shoots; nor while quite young, as the produce may be unsaleable. As a rule both these limits should be determined, as well as the size of the trees which furnish the most useful material. The age of felling, corresponding to that at which the average annual production is greatest, can then be decided with safety. Generally it may be said that, so long as the age at which the trees cease to produce vigorous coppice shoots is not exceeded, the longer the rotation, the more valuable the produce and the greater the revenue.

Period for which the fellings are prescribed.—The fellings should always be prescribed for the whole length of the rotation which is the same as the exploitable age.

Area to be operated on.—The area of the coupes is determined by dividing the total exploitable wooded area by the number of years in the exploitable age. This gives the size of the average coupe which may be increased or diminished as required. Each block or natural sub-division should be divided into an integral number of coupes of approximately the same productive power.

Order to be followed in the fellings.—The rules regarding the allocation of the fellings should be attended to. The coupes should have the most regular form possible, and, as a rule, should succeed one another in consecutive order on the ground; and the produce of a coupe in course of exploitation should not be transported through another coupe recently exploited. A good system of roads or paths must consequently be arranged.

Nature of the fellings.—The sylvicultural rules regarding the fellings to be made are simple. The felling of the trees flush with the ground and at the most suitable season of the year, is, as a rule, all that need be prescribed.

Possibility.—This is prescribed by area, and is determined by fixing upon the area to be operated on. In determining the possibility the present age of the crop as well as its age at the time of felling should, if possible, be stated in order to justify the plan which, owing to the irregularity of the crops, must often be of an abnormal character.

Although the exploitation of the coupes in regular succession, in the order in which they stand on the ground, is desirable, this is not always practicable in the first rotation, especially when dealing with areas which have already been subjected to coppice fellings without the control of a regular working-plan. As the following example indicates, such a case presents no real difficulty in the framing of the plan of fellings.

It is assumed that a plan is being drawn up for a working-circle containing 1,188 acres already worked as coppice and composed as follows in the year 1888-89:—

Block.	Area in acres.				Remarks.
	Wooded.	Blank.	Occupied or unculturable.	Total.	
Chansil	360	3	1	364	Of which 178 aged 12 to 13 years. 182 aged 8 to 9 years.
Kotigall	200	9	...	209	" 8 years.
Datmir	100	1	...	101	" 5 to 6 years.
Naintwar	270	2	3	275	" 4 to 5 "
Sahlra	236	3	...	239	" 130 aged 1 year. 106 " 13 years.
Total	1,166	18	4	1,188	

This analysis enables us to study the composition of the growing stock or capital and to arrive at the following figures:—

Crops	1 to 3 years old						130 acres.
,,	4 to 6	,,	,,				370 ,,
,,	7 to 9	,,	,,				382 ,,
,,	10 to 12	,,	,,	and over			284 ,,

We will further assume that it has been decided to work the forest on a short rotation of twelve years, so as to furnish small fuel for the neighbouring right-holders, and villagers. Consequently, as $\frac{1166}{12}$ acres would be cut over annually, or on an average 291 acres every three years, there ought to be about 290 acres for each of the above age-classes. It is, therefore, apparent that the forest is over rich in woods of medium age, and that the excess capital may be utilized. The average size of each annual coupe would be found, after deducting the blanks, etc., to be $\frac{1166}{12}$ or 97 acres. But we will suppose that in order to provide in the different blocks for differences of fertility which, owing to a complete change of soil, are somewhat marked, it has been decided to distribute the coupes as follows:—

Chansil	—good soil	—4 coupes of	90 acres wooded each.
Kotigall	—medium	—2 ,, of	100 ,, ,, ,,
Datmir	— do.	—1 ,, of	100 ,, ,, ,,
Naintwar	—good soil	—3 ,, of	90 ,, ,, ,,
Sahlra	—bad soil	—2 ,, of	118 ,, ,, ,,

It will also be supposed that advantage has been taken of existing roads and other land-marks in locating the coupes so that the areas are not exactly equal. The fellings to be made are as follows:—

Year of fellings and order of working during first rotation.	No. of coupe.	Block or compartment.	Area in acres.				Age of the crop when felled. Years.	Remarks.
			Wooded.	Blank.	Occupied or uncultivable.	Total.		
1889-90	XII	Sahlra	106	...	...	106	14	As the forest capital is in excess of the normal, coupe No. 12 in Sahlra will be felled in the first year, and again in its turn in the last year of the rotation. This slight departure from the regular order will disappear in the second rotation.
	I	Chansil	92	2	1	95	14 to 15	
1890-91	II	Ditto	86	1	...	87	15 to 16	
1891-92	III	Ditto	94	...	...	94	12 to 13	
1892-93	IV	Ditto	88	...	...	88	13 to 14	
1893-94	V	Kotigall	108	6	...	114	14	
1894-95	VI	Ditto	92	3	...	95	15	
1895-96	VII	Datmir	100	1	...	101	13 to 14	
1896-97	VIII	Naintwar	82	1	...	83	13 to 14	
1897-98	IX	Ditto	96	...	3	99	14 to 15	
1898-99	X	Ditto	92	1	...	93	15 to 16	
1899-1900	XI	Sahlra	130	3	...	133	12	
1900-1901	XII	Ditto	106	...	...	106	11	

It is evident that, if any equal yield were required from year to year, the excess felling rendered 'necessary by the superabundance of mature crops should be spread over several years instead of being carried out in 1889-90.

Conversion of irregular forest into coppice.—In the foregoing example it has been assumed that the forest had already been under coppice treatment, and that consequently there existed on the ground a more or less complete scale of age-classes. Generally speaking, however, this is not the case in India; and the crop to be dealt with not infrequently consists of an irregular and inferior or partially ruined seedling forest or scrub.

In any case the forest must offer one of two conditions. The number of young trees in the growing stock is either sufficient or insufficient in view of reproduction by coppice shoots. In the former case the conversion of the crop into coppice offers no difficulty. It will first be necessary to decide on the age at which the coppice, when created, should be exploited. This, unless there are in the neighbourhood coppice forests of the same kind, must be more or less a matter of guess-work. Having decided on the age, the area to be worked would be divided, in the manner already explained, into as many coupes as there are years in the rotations, and one of these coupes would be clean felled each year.

In dealing, however, with a mature or over-mature crop incapable of being regenerated by coppice shoots it is necessary to regenerate by seed with a view to constituting a new crop which could be converted into coppice while still comparatively young. For this purpose the forest should, as in the former case, be divided into as many coupes as there are years in the age at which it is proposed to exploit the coppice, and the regeneration, either by natural or by artificial means, of one or two of these coupes, should be taken in hand each year. In this way there would in due time be brought into existence a complete scale of age-classes which could then be converted into coppice as in the first example. In both cases cuttings would be prescribed in a table of fellings such as shown at page 70.

Supplementary regulations preserving belts of trees.—It is always useful to preserve belts of trees—which sometimes, for climatic reasons, are of considerable breadth—along the roads, main division lines or boundaries of the simple coppice compartments. Such trees are useful in many ways; they protect the coupes, furnish seed, and, when required, act as boundary marks; and they adorn the forest. The establishment of such belts, the species to be preserved, etc., should therefore be considered, and if necessary should be prescribed in the working plan.

Works of improvement.—The works of improvement required in coppice forests are generally limited to the restocking of blanks, which are common enough in such forests, and in some cases to the construction of ditches or fences for the exclusion of cattle. Coppice forests are not likely to be formed except for the supply of fuel in the immediate neighbourhood of villages and of large towns, and therefore in situations exceedingly liable to trespass. A ditch or a wire fence is often the cheapest way of putting a stop to cattle trespass which, in view of the short intervals at which the stock in coppice forests is renewed, is a grave danger.

COPPICE-SELECTION METHOD.

General plan.—The coppice-selection method of treatment, a selection method in which reproduction is obtained by coppice instead of by seed, is believed to be only applied in India in the treatment of bamboos, of which the "culms" may be compared to coppice shoots. Such fellings may be carried on simultaneously with those under whatever method of treatment is applied to other species in the forest where the bamboos are growing. The whole bamboo-producing area, or so much of it as it is desirable to exploit, may be divided into two or three coupes which are visited in turn every two or three years, as the case may be, care being taken always to leave a certain number of shoots in each clump.

Such fellings are organized and prescribed by area, in the way that would be followed for simple-coppice worked on a very short felling rotation. The possibility regulates itself; and all that is required is to parcel out the area into two or three coupes to be worked in regular rotation.

THE BRANCH-COPPICE METHOD.

General plan.—In certain parts of India, notably in hilly or mountainous country, the inhabitants practise pruning or lopping off branches of trees for firewood and manure or for fodder or litter for their cattle. Where this practice prevails, the transport of timber to a distance is frequently out of the question, and there is often consequently more material than the people can utilize, while fodder during the winter is urgently needed. In such cases the method cannot be condemned, as it is perhaps the sole means of furnishing the fodder required; and it may, therefore, sometimes be necessary to recognise it as a justifiable modification of the coppice method. Conifers, which in the Himalaya are often treated in this way, do not of course throw out shoots like broad-leaved genera; but the smaller branches left and new shoots springing from buds on these branches replace those removed. When necessary, the working of forests according to this method may be organised in the same way as for ordinary coppice on a short rotation of from 5 to 10 years.

Modification of the branch-coppice method.—In the pasture grounds of some countries there is practised a modification of this method which consists in pollarding all the trees at a height of 5 or 6 feet from the ground, so that the young shoots produced may be out of reach of cattle. These shoots are removed a few at a time, as in the coppice-selection method; and the working-plan consists in dividing the area in to a few coupes to be cut over in rotation, and in limiting the size of the branches to be cut. The larger branches bear seed from which a sufficient number of young trees are produced to replace the old pollards as they decay; but reproduction however sought necessitates the exclusion of cattle until the young trees grow out of reach.

METHOD OF COPPICE WITH STANDARDS.

The general plan.—The method of coppice with standards

is applied in exactly the same way as simple coppice, the difference between the two methods—a very great one—being in the selection and reservation of the standards. The exploitable age is calculated for the underwood only, and in the same way as in the case of simple coppice. Generally, however, the underwood is felled at a more advanced age, as this procedure tends to lengthen the stems of the reserves and has other advantages. The possibility is prescribed, as in simple coppice, by area, but with the addition of a rule regulating the constitution of the reserve of standards: that is to say, prescribing the number of stems of each species and of one rotation which must be reserved at each exploitation, and specifying the number of older stems to be felled.

Reservation of standards.—The characteristic of the method lies in the reservation of the standards, and the value and exact constitution of this reserve must be determined with great care. The number and proportion of each class of standards have, therefore, to be decided. The total number of all classes that can be retained is limited by the fact that standards should remain isolated after they are first marked and until they are felled. The *maximum* number per acre is, therefore, the area of one acre divided by the average area covered by the crown of one mature standard; but the number that is reserved in practice depends on the species both in the reserve and in the coppice. As regeneration is principally obtained by means of coppice, the cover of the standard trees must not be of a kind to unduly interfere with the development of the stool shoots. Provided no such interference occurs, the greater the number of standards the better; as this reserve enormously increases both the capital value and the revenue. The value of the coppice, as compared with that of a fully established reserve of standards, is insignificant.

The following example exhibits the method by which the number of standards can be arrived at, and the influence of the reservation on the revenue and capital value of the forest. It will be assumed that the length of the rotation is 20 years; that the maximum age up to which the standards can be preserved is 80 years; and that it has been found by experiment that the *cover* of the trees of different ages is as follows:—

		Square feet.
Trees of one rotation or 20 years, each	. . . .	30
,, two ,, 40 ,,	. . . .	150
,, three ,, 60 ,,	. . . .	400
,, four ,, 80 ,,	. . . .	600

The number of square feet in one acre is 43,560; and as $\frac{43,560}{6.0} = 72$, a crop of that number of trees of 80 years old would form a *complete crop* on one acre. Let us suppose that, in view of this and from observation on the spot, it is decided that the total number of standards per acre should not exceed about 40 trees, and that trees of from 60 to 80 years old have attained the most useful dimensions they can reach while sound; also that it has been estimated that only about a third of the stems first reserved can be again reserved with advantage at the second rotation, and so as regards these again. There should, therefore, be reserved at each exploitation.—

    27 standards of one rotation.
     9   "   " two   "
     3   "   " three "

As soon as the capital was constituted the number of standards to be felled at each exploitation would, therefore, be:—

  Trees of one rotation, felled 0, reserved . . 27
    "   " two   "   " 18,   "   . . 9
    "   " three "   " 6,   "   . . 3
    "   " four "   " 3,   "   . . 0

When the capital was fully constituted and just before each coupe there would be per acre (in addition to the coppice containing 27 standards of one rotation about to be reserved and covering an area of 810 square feet):—

                                                                     Square feet.
  27 trees of two rotations covering 4,050
   9   "   " three   "   "   3,600
   3   "   " four   "   "   1,800

                    TOTAL AREA COVERED . 9,450

The maximum area covered by the standards, including that covered by the trees about to be reserved, would, therefore, only amount to one-fourth of the total area. For the particular species concerned it would, by hypothesis, have already been ascertained that the room left for the coppice growth is sufficient.

As regards the effect of the reservation on the capital value of the forest and on the revenue, we may assume the net value of the tree of 40 years old to be R5, of 60 years old R10, and of 80 years old R20; and that an acre completely stocked with simple coppice 20 years old produces at each felling a net profit of R100. The effect of the reservation of the standards would therefore be to raise the net value of each one-acre coupe when the crop is mature from R100 to R286, and the net *annual revenue* per acre from R5 to R14¼. Capitalised at 4 per cent the value of the forest cropped with simple coppice would be R125 per acre; while with standards, the value would be R360. The results may in many cases be even more favourable to compound coppice.

There would be felled at each coupe.—
                         R   R
18 trees, 40 years old, @ 5 = 90
 6   "   60   "   "   10 = 60
 3   "   80   "   "   20 = 60
                                 210
Coppice: R100 − $\frac{100 \times 10,260}{43,560} = 76$

             TOTAL . 286

In selecting the standards attention is paid to their origin, their species, their shape, their condition of growth and their position in the crop. Seedling poles should be preferred to coppice, because they are longer lived. Where, however, seedlings are not found, sounds shoots from young and small stools may be chosen. Only such species as furnish valuable timber should be reserved; inferior trees should not be selected. It is not necessary to reserve the prescribed number of standards in every acre or coupe : only well-shapen, straight, sound and vigourous stems should be chosen, and they should be selected in places where it will be possible for them to thrive. This should all be prescribed in a regulation attached to the statement of fellings.

Supplementary regulations.—The importance of a numerous and well-constituted reserve, formed chiefly of stems which have sprung from seed, has been explained. There is, however, a likelihood of the seedlings not being forthcoming owing to their being suppressed by the faster growing coppice. Means are, therefore, sometimes taken to foster them by cleanings during the early years of their existence. Such cleanings are frequently conducted—every year or every two or three years according to the species and the rate of growth—until the coppice has attained a certain age, 5, 10 or 15 years, when they are as a rule discontinued. In certain instances it may be advisable to make a thinning 2 or 3 years before the coppice is felled, in view to the better development of such stems as will probably be selected for standards.

Conversion of irregular high forest into coppice with standards.—This conversion, except as regards reserving standards, can be done in exactly the same manner as has been explained with regard to simple coppice. If there is a sufficiency of young trees capable of producing coppice shoots, the conversion can be commenced at once, the forest being divided into coupes to be felled in rotation, just as if these coupes already contained coppice growth. Otherwise it will be necessary first to constitute young crops which can then be operated on as in the former case. The young growth can either be obtained naturally by regeneration fellings or artificially by sowing or planting. The reservation of the standards obviously presents no difficulty. The stems it is wished to retain are immediately selected, and the reserve constituted. Even where the crop is so mature that it is necessary, before applying the new method, to obtain a young growth, trees should be maintained with a view to constituting the reserve with the desired proportion of older stems when the coppice fellings can be begun. The possibility, which will be by area for coppice conversions when undertaken direct, may be calculated in the same way as already stated for ordinary coppice fellings.

Transformation of simple coppice into coppice with standards.—The operation of converting simple coppice into coppice with standards is very simple; but, in order to produce the desired capital, it has to be spread over as many rotations as are necessary to obtain standards of the oldest age desired. It merely consists in the reservation of the required number and kind of standards in each coupe. Beyond prescribing the reservation of standards, no change in the working-plan would be required. The fellings would otherwise be organized as already explained.

METHOD OF CLEARANCES.

General plan.—This method, as already explained, includes several different forms, viz., clearings on adjacent areas, clearings on alternate parallel strips, and clean fellings with artificial regeneration. In all these modifications the possibility is determined by area, as in the method of simple coppice, and the number of coupes, if annual, is in each case made equal to the number of years in the exploitable age; if biennial, to half the number of years, etc. The fellings are, however, prescribed with some slight differences in each of the modifications.

In the system of clearings on adjacent area, the exploitable age having been determined, the whole area is divided into as many coupes of equal fertility or equal resources as there are years in the exploitable age, or, if necessary, into one-half or one-third as many biennial or triennial coupes, one of which would be felled every year or every two or three years, as the case might be. The clearances made may be either clear fellings, or a certain number of trees may be reserved to grow to a larger size and to assist in the regeneration by the seed they shed. Where no trees are reserved natural regeneration by seed can only be secured from seed shed by adjacent trees, and the average area of the clearances should, therefore, be small.

The object of the strip system, which is the same in principle as that of the method of adjacent areas, is to ensure natural regeneration taking place over large areas. With this object long narrow strips are marked out on the ground, and every alternate strip is cleared until the whole area has been traversed, when the alternate strips, omitted at the first passage of the cuttings, are clean felled in their turn. By these means a newly felled coupe has, as it were, a hedge of seed bearing and sheltering trees on either side of it. Larger areas may obviously be felled without risking failure of reproduction than in the case of the method of adjacent areas, and as in that method reserves may be left if desirable.

The **system of clean-fellings (with artificial regeneration)** is applied in the same way as the method of clearings, except that, instead of trusting to reproduction from seed falling from the reserved trees or from the adjacent forest, the area felled is planted up or sown each year. The size of the coupes may obviously be as large as can conveniently be re-stocked artificially; and, unless re-stocking fails, the ideal or normal type of forest may be approached very closely under this method. The cost and difficulty of successfully re-stocking large areas are the chief drawbacks to its employment.

METHOD OF STOREYED FOREST.

General plan.—In applying this method the number of stems of each age or size-class to be reserved must be decided on. This number is deduced from the area covered by the crown of the average tree of each size-class and from the rate of growth. The difference in age between each class is, we may assume, equal to the length of the felling rotation; and the classes must be formed so that the trees of one class will attain the dimensions of the next higher class during that interval. Or, if preferable, the size-classes may first be decided on, and the length of the felling rotation, based on the rate of growth, may be made to correspond.

Thus, suppose that the diameters of the size-classes determined on are: below ¼', ¼' to 1,' 1' to 1¼', 1½' to 2', and over 2 feet; that trees of 2 feet diameter have attained their maximum utility; and that the rate of growth is such that in about 30 years the trees of the lowest dimension in one class attain the minimum size for the next higher class. This period of 30 years would, therefore, be taken as the length of the felling rotation. In such a forest the capital, when normally constituted would consist of four age-classes, each occupying one-fourth the total area. Thus in one acre, containing 43,560 square feet, each age-class would cover one-fourth of this area, or 10,890 square feet. There would be no crops composed of trees over 2 feet in diameter, as in theory these would be exploited as soon as they reached that size:—

Class I, stems over 2 feet diameter, occupying . .	Nil.
„ II, „ „ 1½ to 2 feet diameter, occupying .	10,890 square feet
„ III, „ „ 1 „ 1½ „ „ „ . .	10,890 „ „
„ IV, „ „ ½ „ 1 foot „ „ . .	10,890 „ „
„ V, „ below ½ foot	10,890 „ „
TOTAL .	43,560

In order to ascertain the number of stems of each class, it would be necessary to measure the areas covered by the crowns of the average trees in each class. We will assume these areas to be as follows :—

Class I	900 square feet.
„ II	625 „ „
„ III	400 „ „
„ IV	100 „ „
„ V (dominant stems)	25 „ „

The number of trees of each class, when the capital has been constituted, is evidently the area covered by the class (10,890 square feet), divided by the average area covered by one crown of that class; and would, therefore, be as follows :—

Class I	0 trees.
„ II	17 „
„ III	27 „
„ IV	108 „
„ V	436 „

In the course of 30 years, according to the assumption made, all the trees in each class attain the next higher class. If, therefore, during this interval we fell the difference in numbers the capital will remain unaltered. This operation may be tabulated as follows:—

CLASS.	Felled.	Reserved.	TOTAL.
I	17	...	17
II	10	17	27
III	81	27	108
IV	328	108	436

At the end of the thirty-year period the crop on the ground would thus be the normal capital. The place of the young growth in Class V at the commencement of the period would be taken by a new stock of seedlings which would occupy one-fourth of the area, i. e., the portion not covered by the standards reserved.

This method is applied in the same way as the selection method. The forest is divided into whatever number of annual or periodic coupes the felling rotation requires, and the number of trees to be felled each year is prescribed. The number of stems to be felled in each size or age-class should be stated.

When first applying such a method to a forest, the number of stems to be felled would not of course, as in the example given, be the difference between the numbers of trees in the various classes. In such a case, the number of stems *to be left* on the ground having been determined in the manner explained, reference would be made to the results of the enumeration, and the number of trees to be felled would be deduced therefrom.

Thus, suppose an enumeration of the block or compartment to be exploited, containing 200 acres, gave the following results:—

Class	I	.	.	470 trees;	or	.	.	2 trees per acre.
„	II	.	.	5,656 „	„	.	.	28 „ „
„	III	.	.	3,388 „	„	.	.	17 „ „
„	IV	.	.	26,690 „	„	.	.	133 „ „
„	V	.	.	127,400 „	„	.	.	637 „ „

In order that the capital should be constituted in the manner required, it would be necessary at the first operation to fell and reserve, on an average per acre of the coupe:—

Of Class	I trees;	to fell	2 trees;	reserve	.	0 trees.
„	II „	„	11 „	„	.	17 „
„	III „	„	0 „	„	.	17 „
„	IV „	„	15 „	„	·	118 „

There would be reserved of Class IV 118 trees instead of 108, to make up for the deficient number of trees reserved in Class III.

It need hardly be said that, when a mixed crop is dealt with, the felling of each species should not be separately prescribed as has been sometimes done in Indian working-plans. At most the relative proportion of each kind should be prescribed. This "storeyed forest" method of treatment has been seldom described, though it has been largely applied in certain parts of France. It is stated to be particularly well-suited to hill forests, in which the uncovering of the soil leads to a dense growth of forest weeds and thus prevents reproduction.

THE METHOD OF SELECTION FELLINGS (FRENCH *JARDINAGE*).

The General plan.—The method of selection fellings consists in felling here and there, as they are found growing and according to certain cultural rules, such of the exploitable trees as it is calculated will not exceed the possibility of the forest. Theoretically, every acre of a selection-worked forest should be felled over each year; but, as the trees felled would be so scattered that it would often not be profitable to remove them, it is usual to fell over a portion of the area each year and thus work over the whole area in a given period at the end of which the portion first worked is again taken in hand. The working-plan thus prescribes the number of years in which the whole area is to be worked over; the area which is to be worked each year, or during a short period of years; and it limits the quantity of material to be removed, annually or periodically, to the total production over the whole area during the year or period concerned.

Thus the working-plan might be to the effect that each year there will be felled by the selection method, on one-tenth of the area of the forest, such and such a number of trees.

Limitation of the fellings.—The limitation of the material to be removed may be effected in several ways, *viz.* :—

(*a*) by cultural rules only ;

(*b*) by prescribing, in addition to cultural rules, the quantity of material or the girth or diameter limits of the trees to be felled ; and

(*c*) by prescribing the proportion which the number of trees felled shall bear to the number of stems of certain dimensions in the standing stock. The number of trees or the volume of material which is to be felled may, when precision is required, be calculated by one or other of the following methods which will be separately described in the order given :—

      The rate of growth.
      Enumeration of the trees.
      Estimation of the production of the soil.
      The method of relative proportion.
      The method of proportionate volumes.

Another process based on tables showing the yield of different classes of forest and soil need not here be dealt with, as no such tables for Indian forests exist.

It may be useful to note that in applying these methods, the yield need not as is frequently done be separately calculated for each species. The enumeration would show the relative proportion of each species; and in the working-plan the fellings of each may be prescribed according to that proportion. Thus, suppose that ⅓ of the trees enumerated were of species A, and ⅔ of species B, and that the possibility were fixed at, say, 600 trees a year, the plan might prescribe the felling of 200 trees of species A and 400 of species B.

Fellings limited by cultural rules.—The cultural rules prescribing the method of making the fellings should be easy to understand and of general application; and they should also ensure that, as far as possible, more material than the forest produces will not be removed. Where the demand and consequently the fellings are very light nothing more is required than to fix the diameter below which trees must not be felled, or to limit the fellings to the removal, here and there, according to the principles of the selection method, of such trees as are over-mature or are above a certain girth.

Such general rules sufficiently limit the fellings where the crop is already constituted according to the selection type or where there is a good executive staff. But where, as often happens, this is not the case, there is danger of such simple rules being unintelligently or unscrupulously applied.

To meet this difficulty it is well, when dealing with large irregular forest masses, to supplement, as in the following example, the general rules by hints or directions, conveyed in the "Remarks" column of the description of each block, regarding the nature of the fellings to be made:—

Description of Compartments.

Name of block.	Area in acres.				Configuration and aspect.	Soil.	Stock.	Remarks.
	Wooded.	Grass blanks.	Unproductive ground.	Total area.				Suggestions as to future treatment, etc.
Mattiungra.	102	32	25	159	Intersected by numerous narrow-topped ridges. All aspects are represented. Gradients 28°—30°. Altitude 6,000 to 6,300 feet.	Granitic. Sandy clay; very deep.	About one-fifth open grassy blanks, with a few mature ban, rai, laurl, etc. Stocked portions contain mature oaks and kokat, with a few rai, ban, ayar, and burans, except in nalas where some saplings are found. Nearly all old trees unsound.	The mature trees on the edges of blanks, as well as those on the open grassy slopes, should be left as seed-bearers; and the few trees which can be taken out should only be felled along nalas. Grass blanks might be planted up.

Cases in which the fellings may properly be regulated by cultural rules.—The cultural method of organizing selection-worked forests is especially applicable to the irregular and partially-ruined condition so frequently to be dealt with in the forests of India, where enumeration surveys would often be waste of time and money. The method is also applicable to forests in which it is merely desired to retain the cover. In the latter case, the fellings would, of course, be limited to the removal of dead or decaying trees.

But the method enforced in the manner explained above is attended with the drawback that there is no means, other than by personal inspection, of checking its correct application. The only extraneous control that can be exercised over its application is with regard to the *area* exploited; and all the prescriptions on this subject might be rigidly adhered to while the far more important cultural rules were being misapplied by unintelligent or unscrupulous subordinate officials. Hence, where possible, it is always preferable to determine the quantity of material that may annually or periodically be removed with safety, and to limit the fellings to this maximum quantity while subordinating them to cultural rules.

Thus, suppose it were ascertained by experiment that the average production of a crop in its present condition was 30 to 40 cubic feet per acre per annum, and that it was desirable to increase the forest capital, the fellings might be limited to 20 cubic feet per acre per annum on an average for the whole area, or, if the area were 1,000 acres, to 20,000 cubic feet. This maximum amount could be felled in each annual coupe.

Limitation of fellings determined by the rate of growth.—It was explained, when discussing at pages 5 and 6, the formation of the forest capital, that, in a normal forest, the number of trees which attain exploitable dimensions in a given period practically represents the possibility of the forest for that period. As it is possible to estimate, for a short period in advance, the number of trees in a forest that will become exploitable and to determine with some accuracy whether the crop approaches the normal type or not, the principle may, in some cases, be usefully applied to selection-working in forests in which the trees can be enumerated.

Suppose a forest in which trees of all ages are well represented and fairly evenly distributed. and in which the rate of growth is such that in the course of 30 years trees of $4\frac{1}{2}$ feet girth attain the minimum exploitable girth-limit of 6 feet. In the course of 30 years all the trees now $4\frac{1}{2}$ to 6 feet in girth would be removed. Provided, therefore, that trees of all sizes now from $4\frac{1}{2}$ to 6 feet girth were properly represented in the crop, and were felled as they attained to 6 feet girth, the annual possibility would theoretically amount to one-thirtieth of the total number of trees above 4½ feet in girth standing in the forest. In practice since the exploitable size is a minimum only and since merely a portion of the forest is gone over annually, a stock of exploitable trees has to be accumulated and the possibility is necessarily something less.

As an adjunct to cultural rules, and subject to such restrictions as common sense and a knowledge of sylviculture dictate, the principle here sketched may often be applied with advantage. It may even, with some sacrifice of accuracy, be used to limit the fellings in forests in which, for any reason, it is inexpedient or impossible to carry out *complete* enumeration surveys in advance of the working.

In such cases, all that is necessary is to determine the average rate of growth of the principal species when approaching maturity and the length of the felling rotation, and to prescribe, by a simple rule, the proportion of the trees above certain dimensions growing in each coupe that may be removed in each felling.

Thus, suppose that for the forest dealt with in the last example it were determined to work over the whole area in ten years by annual coupes of one-tenth of the area. Each of these coupes, it might be assumed, would contain about one-tenth of the trees growing in the whole area, so that to fell one-third (that is ten times one-thirtieth) of the exploitable trees in a coupe would be equivalent (as regards the number of trees felled) to the felling over the entire forest of one-thirtieth of the stems exceeding a certain prescribed minimum girth. This, however, pre-supposes that the accumulation of trees 6 feet in girth and over is such that the requisite number is available on one-tenth of the area, and does not justify the felling of trees over 4½ but less than six feet in girth. A rule, therefore, prescribing that *there will be felled each year on one-tenth of the area, in succession, one-third of the total number of trees of from 4½ to 6 feet in girth growing on that area*, provided no tree is felled before it attains the exploitable size, would in the restriction of fellings, have much the same effect as if the whole of the trees in the forest were counted in advance and the exact number to be felled each year were prescribed.

It is unnecessary to discuss in detail either the errors involved in this method or the cases in which it is applicable. There are undoubtedly many instances in which the method might be applied, and in which it would lead to better results than any attempt, with the means usually available in India, to enumerate in advance the whole stock.

Limitation of fellings derived from an enumeration of the trees.—The principle explained in the preceding paragraph may also, it is obvious, be applied to the results of an enumeration of the growing stock made in advance of the exploitation. The number of trees of each kind and size-class and their rates of growth being known, it is possible to estimate with fair accuracy the number that will pass from one size-class to the next higher class, or that will become exploitable in a given period of time. The average number of trees that will be available annually for felling during this period is thus known.

As an example, suppose a forest at present containing the following stock of exploitable trees, and in which the average annual diametral increase of the stems when approaching the exploitable size of 2 feet in diameter is 0·2 inches—

Class	I	over 24 inches in diameter	.	.	24,741 trees.
,,	II 18 inches to 24	,,	,,	.	17,867 ,,
,,	III 12 ,, 18	,,	,,	.	19,467 ,,
,,	IV 6 ,, 12	,,	,,	.	192,827 ,,
,,	V below 6	,,	,,	.	968,000 ,,

In the course of 30 years the growth in diameter of the larger trees will be $0·2 \times 30 = 6·0$ inches. Consequently in that time the trees of Class II, now 18 inches to 24 inches in diameter, will be replaced by those in Class III : they may therefore be removed. Most of them will have attained the exploitable diameter, but a certain number being suppressed or crowded out may, unless felled while still below the exploitable size, be ultimately unutilisable. Assuming, therefore, the stock to be complete and normal, we may theoretically fell in the course of 30 years, without exceeding the possibility, all the trees now in Class I and II, that is to say, $24,741 + 17,867 = 42,608$ trees or at the rate of $\frac{42,608}{30} = 1,420$ trees a year.

In the preceding remarks it has been assumed that *the crop contains a sufficiency of trees of the lower age-classes.* But in a forest where, for example, most of the trees are mature or are approaching maturity—and this is the condition of crops which have not been worked or have been much under-worked—fellings determined in this manner would remove in a single period not the possibility but practically the whole forest capital.

In India it has sometimes been the practice to base the calculation on the number of trees already exploitable, and to limit the annual fellings to this number divided by the number of years in which all the stems in the next lower size-class will become exploitable. Thus, in the above example, the annual fellings might be limited to 24,741 divided by 30, or to 825 trees.

In other cases, where the younger age-classes are sufficiently well represented, the possibility has been arrived at by dividing the exploitable trees together with a proportion of those to become exploitable during the felling rotation by the number of years in that rotation.

The calculation can, however, be made in a less rude fashion. Supposing that the ages are evenly graduated, the average number of trees that can attain exploitable dimensions each year in the immediate future is the total number in Class II (the size-class next below that of which all the trees are exploitable) divided by the number of years required for trees of the lowest dimensions of this class to become exploitable. This is then the possibility of the crop for the time being; and it must be estimated—from the average production of other similar forests for instance—whether the result so obtained is above or below the normal or potential possibility, and the number of trees to be felled should be increased or diminished accordingly. With a view to estimating whether the age-classes are suitably graduated or not, the number of trees in Class II may be compared with those in the lower and higher classes. Finally, as regards the sufficiency or insufficiency of the stock already exploitable, it must be remembered that, in a normal forest in which the age-classes occupied equal areas, there would be no trees above the exploitable size on the ground immediately after felling, and only one year's growth immediately before the next felling. In a selection-worked forest, owing to the trees of different ages being distributed all over the area, this would only be the case where the whole area of the forest was worked over each year. Ordinarily, however, as already explained, the entire forest is gone over in a number of years; and consequently it is only on the portion of the area felled over twelve months previously that there is one year's growth of exploitable trees, on the next area there is two years' growth, on the next three, and so on up to the limit of the felling rotation. Knowing, therefore, the number of trees of Class II which annually attain exploitable dimensions, we can calculate the normal exploitable stock and thus ascertain whether the actual exploitable stock is sufficient, insufficient or superabundant relatively to the stock in Class II. If superabundant, the excess can be utilised at once or in several years according as the lower stages of growth are sufficient or insufficient. If insufficient, less than the normal possibility should be removed.

The crop already taken as an example will serve to illustrate this method of analysis. The number of trees below exploitable dimensions, 18 to 24 inches in diameter, being 17,867, the greatest number of trees (the 24,741 trees over 24 inches being already exploitable) which can attain exploitable dimensions each year in the immediate future is $17,867 \div 30 = 596$ trees. This number, divided by the wooded area of the forest expressed in acres, would give the average annual production of exploitable trees per acre. Assuming this area to be 1,200 acres, the average annual production per acre would be $596 \div 1,200 = 0.5$ trees. It would be possible to compare this figure with that of other similar forests, and thus to ascertain whether the production was above or below what it ought to be. We will suppose that it is found to be considerably in defect. It would follow that the forest is deficient in trees of Class II, and also it would appear of Class III, as the number is about the same. Class IV would, however, appear (a rough approximation only is possible) to be well represented, and so would Class V. It may, therefore, be accepted that during (we will assume) the next 60 years, until the crops of Class IV begin to be exploitable, the fellings must be made with caution; but that the crop will be thereafter fairly complete if not altogether normal. This interval of 60 years is in fact a preparatory period during which a provisional plan is required.

As regards the sufficiency or otherwise of the existing exploitable stock, we will suppose the felling rotation adopted to be a very long, say 30 years, and that the whole area is sub-divided into 30 portions each approximately equal in extent. Every year there would pass from Class II into Class I and become exploitable something less than 596 trees, or on one-thirtieth of the area $\frac{596}{30} = 19$ trees. Consequently immediately before the commencement of the second felling rotation and ever after, the exploitable stock on the ground would be:—

On the area felled over 30 years before 30×19 trees.
,, ,, ,, 29 ,, 29×19 ,,
,, ,, ,, 28 ,, 28×19 ,,
etc., ,, ,, etc., etc.,
,, ,, ,, 2 ,, 2×19 ,,
,, ,, ,, 1 ,, 1×19 ,,

The total number of exploitable trees left standing would therefore be—

$$19 \times (30 + 29 + \ldots + 2 + 1) = 8,835 \text{ trees.}$$

Therefore the surplus stock is $24,741 - 8,835 = 15,906$ trees.

As, however, the crop is defective in trees of Classes II and III, this surplus ought, if the trees can be preserved in a healthy condition, to be utilised during the whole of the preparatory period of 60 years. It would, therefore, be only permissible to fell $15,996 \div 60 = 265$ trees of the surplus stock each year or, in all, $265 + 596 = 861$ as a maximum each year.

The foregoing discussion, as to the sufficiency of the stock already exploitable, proceeds on the assumption that the stock of trees in Class II is normal. But suppose the maximum number of trees becoming exploitable each year, viz., 596, is below the normal production which should annually be about one tree an acre or 1,200 trees in all. In this case it would be desirable to preserve on the ground a corresponding stock which would be, with a felling rotation of 30 years, $\frac{1,200}{30}\{30 + 29 + \ldots 1\} = \frac{1,200}{30}\{30 + 1\}\frac{30}{2} = 18,600$. Here again it would only be permissible to fell $(24,741 - 18,600) \div 60 = 102$ trees of the surplus stock each year during the preparatory period; so that the fellings would be reduced to $102 + 596 =$ about 700 trees a year. Again it might be that the exploitable stock is *deficient*. If there were only, we will suppose, 4,741 instead of 24,741 exploitable trees, and the stock required were, as in the last case, 18,600 trees, the forest would be very deficient in mature stock. In the course of 60 years, the exploitable stock should be augmented by the difference $18,600 - 4,741 = 13,859$ trees, or at the rate $\frac{13,859}{60} = 231$ trees a year. In such a case, the number of trees in Class II remaining the same as before, there ought not to be felled each year more than about $596 - 231 = 365$ trees.

Fellings limited by the productive capacity of the soil.—The method last described meets the difficulty, so frequent in India, involved in calculating the possibility of a mixed crop containing only one or two saleable species, such as the teak forests of Burma, the deodar forests of the Himalaya, and many others. The more accurate methods employed in European forestry are not intended for such forests, and it is a quetsion whether they can be made use of except where the whole crop is saleable.

The most common of these European methods consists in determining the number of trees to be felled by estimating the production of the soil. The number is, if possible, determined by ascertaining the mean increment of a complete crop of the exploitable age on the same kind of soil, and by dividing this increment by the volume of an average tree of the exploitable size.

Thus, if the average annual production of the soil were 120 cubic feet per acre and the volume of an average exploitable tree 60 cubic feet, the possibility would be fixed at two trees per acre per annum.

As, however, each portion of the forest might not be able to furnish per acre the number of exploitable trees so determined, no minimum girth limit is fixed with regard to the trees to be felled. In this way, a quantity of material less than the possibility is removed from those portions of the forest which are deficient in large trees; while, where the exploitable stems are in excess, the possibility is, for the time being, exceeded. The result is the ultimate regularisation of the crop as a whole.

The method may best be explained by an example. Let us suppose that the forest which it is wished to organize by the selection method, and the possibility of which has to be ascertained, consists of 1,810 acres of mixed conifers and oaks, all being saleable. We will also suppose that an enumeration of the trees has been made and that the forest has been found to contain 263,520 stems of all sizes and ages, the proportion between the two species being 90 firs to 10 oaks ; also that the average number of stems of all ages per acre is 146 as follows :—

Small trees under 12" in diameter	92
Medium-sized trees 12" to 24" in diameter	46
Exploitable trees over 24" in diameter	8
Total stems per acre	146

The production of the soil can be roughly estimated as follows. It may be assumed that the same kind of soil, when fully cropped with trees of a certain species, produces a constant quantity of material per unit of area each year. Suppose that a fir tree of two feet in diameter is exploitable, and that by experiments in the forests it has been shown that a tree of this size requires a circular space of 25 feet across. We will also assume that it has been determined that the mean volume of wood in a typical fir of the above dimensions is 135 cubic feet, and that the tree requires 150 years to attain the diameter stated. The productive capacity or capability of the soil will be as follows :—

The number of trees of the exploitable size that can grow on one acre will be $= \frac{\text{square feet in one acre}}{25^2} = \frac{43,500}{625} = 70$ trees, and the volume of wood will be $70 \times 135 = 9,450$ cubic feet. This is the growth of 150 years. The average mean annual production of each acre, therefore, amounts to $\frac{9,450}{150} = 63$ cubic feet.

In a selection-worked forest the rate of growth is exceedingly difficult, we may say impossible, to determine accurately, because of the inequalities in the growth of the trees at different stages of their existence. But this is no reason for not attempting to estimate the growth, making due allowances for probable error. It is hardly necessary to state that if there existed in the neighbourhood of the selection-worked forest a mature regular high forest of the same species and growing under similar conditions, the productive capacity of the soil could be determined more easily and with greater accuracy by felling and cubing the wood upon a known area. Moreover, it would be possible to ascertain, from a regular crop of the kind, the number of exploitable trees that can grow at the same time on one acre.

Returning to the forest cited as an example, the area of the blanks, roads, occupied lands, etc., has been ascertained by survey to be 210 acres. The productive *wooded surface* would thus be reduced to 160 acres, and the average annual production of the whole area would then be $1,600 \times 63$ cubic feet $= 100,800$ cubic feet. We might stop here and prescribe the removal annually of this volume—which is by assumption equal to the annual production—from the entire area, trusting that the forest would produce the same quantity and thus repair the loss. And, if the forest were perfectly constituted and contained in the necessary proportions trees of every size and stage of growth, no harm would result from this procedure. But such a state of things is not often found, and it is impossible to say exactly how far the actual approaches to the ideal condition; whether there are, for instance, a sufficient number of trees of 2 feet in diameter to furnish 100,800 cubic feet of wood *each year* during the time that is required for the trees of the next (say 1½ feet diameter) stage to attain the exploitable dimensions; that the trees of the next lower stage are sufficient in their turn; and so on.

The yield must, therefore, be fixed in *number of trees*. The annual production of the soil being 63 cubic feet per acre, and an exploitable fir tree of 2 feet in diameter having a volume of 135 cubic feet, the annual yield of each acre in trees will be $63 \div 135 = 0.47$, and, for the entire wooded area, the annual yield will be $0.47 \times 1,600 = 752$ exploitable trees.

It would usually be impossible, without great loss and inconvenience, to spread the fellings each year over the entire area of the forest. We will suppose, therefore, that the annual fellings are limited to one-tenth of the entire area, and that ten years are allowed to elapse between successive fellings in the same area. As we have assumed that there are in every 100 trees 90 firs and 10 oaks, the provisions of the working-plan as regards the fellings might be summed up as follows:—

"Each year there will be felled by the selection method, on successive annual coupes of one-tenth the area of the forest, 752 trees, *viz.*, 677 firs and 75 oaks."

No limit as to the size of the trees felled need be fixed; but in each coupe the dead or dying trees are felled first, and then a sufficiency of the largest trees to complete the prescribed number, which in no case should be exceeded. This system of not fixing any limit to the size, while strictly limiting the number, is characteristic of the method by the application of which the crops, in the poorer portions of the forest, will become richer. Where the trees are small, the same number of stems will contain a less volume of wood; while, where the crop is rich in large trees, a greater volume than the normal production will be felled. The crop will tend to become uniform throughout, and to assume the normal state of one exploited at an average age of 150 years, which is the number of years that the trees require in order to attain 2 feet in diameter, or the assumed exploitable size. There will usually be no danger of all the large trees being felled at once, because all will not be found in one portion of the crop.

An examination of the valuation survey records of the number of trees in each size-class will show the average age or size at which trees have been exploited; for trees above this size will be rare. The record will also readily indicate whether or not there are trees of the various size-classes in numbers sufficient, as far as can be judged, to replace the mature trees when felled, so as to enable the forest always to remain in more or less the same condition as regards its capability of furnishing a given number of trees.

Fellings limited by relative proportion.—This method is based on considerations similar to those which influence the determination of the possibility of regular high forests, composed of properly graduated series of crops of all ages up to the number of years comprised in the normal rotation. These crops may be classed under one of three groups, each of which would obviously occupy one-third of the total area, *viz.* :—

> I.—Full-aged crops, of which the ages will range downwards to two-thirds of the total number of years comprised in the exploitable age.
> II.—Medium-aged crops, of which the ages will range from two-thirds to one-third the number of years in that age.
> III.—Young crops, aged less than one-third the number of years in the exploitable age.

It is evident that, in a regular high forest so stocked and in which the elements of production are everywhere identical, a sustained yield will be assured if each group is in its turn regenerated during one-third the number of years in the exploitable age. The annual yield can, therefore, be determined by dividing the volume of material contained in the full-aged crops, with the addition, if possible, of its increment up to the time of felling, by the number of years in one-third of the exploitable age. When a forest does not contain crops of all ages covering equal areas, it may be necessary to make transfers from one group to another, so that the areas to be exploited in equal intervals may be equalised.

If for a regular be substituted a selection-worked high forest, it is still possible, notwithstanding the apparent irregularity of the latter, to distinguish in it these three groups, namely, of full-aged, medium-aged and young crops. But the areas respectively stocked with these three groups, instead of forming compact blocks, are scattered and intermixed in the most irregular manner throughout the forest. It is usually impossible to ascertain what area each group occupies ; so that it is necessary to determine whether the volumes of material in each group are in normal proportion to each other. It has been ascertained that, in a normally-stocked high forest divided as explained above into three age-groups, the volume of material in the crops of the second group is equal to three-fifths of the volume of the crops in the first group.*

Therefore, whenever in a selection-worked forest the volume of the material in the group of full-aged crops exceeds by two-thirds the volume of the group of medium-aged crops, it may be admitted that these two groups correspond to the two similar age-groups of a normally-stocked high forest; and if it is arranged to exploit the group of full-aged trees in one-third of the number of years corresponding to the age of the exploitable tree, there should be no decrease in the yield during the following period when the trees which, at the outset, constituted the group of medium-aged crops will be felled. Furthermore, all risk of overfelling may be avoided by omitting to take into account the increment of the full-aged crops during the period prescribed for their exploitation.

In practice it is convenient to base the classification on some more easily ascertainable factor, such as the circumference or diameter of the trees, and thus prescribe the maximum size for trees in the groups of medium-aged and young crops. Usually the operation can be still further simplified by determining the size corresponding to the age of exploitability, and by assuming that the full-aged and medium-aged crops comprise respectively trees exceeding two-thirds the size of the exploitable tree, and trees from one to two-thirds of that size. The method is very easy to apply in calculating the yield of selection-worked forests. First, the age of exploitability and the circumference or diameter corresponding to it should be determined. Next, a valuation survey should be made and the trees and their volumes classed as follows:—

    I.—In the group of full-aged crops—when they are more than two-thirds the size of the exploitable tree.

---

* That this is very approximately correct may be seen from fig. 1 by actually counting the rectangles representing the production, or in a similar figure by assuming the average annual growth per acre to be uniform and equal from year to year. If a be the area, v the average annual production per unit of area, and n the number of years in the exploitable age, the value of wood in the first group would be—

$$\frac{\frac{2}{3}nv + nv}{2} \times \frac{a}{3} \text{ and in the second group } \frac{\frac{2}{3}nv + \frac{1}{3}nv}{2} \times a$$

$$\text{Hence } \frac{\text{Volume of wood in first group}}{\text{Volume of wood in second group}} = \frac{\frac{2}{3}+1}{\frac{2}{3}+\frac{1}{3}} = \frac{5}{3}.$$

II.—In the group of medium-aged crops—when they are between one and two-thirds that size.

There is usually no need to count the trees which should be placed in the group of young crops.

If, when the classification is effected, it is found that the total volume of trees in Class I is greater by two-thirds than the total volume in Class II, it may be assumed that the crops are in the normal proportion. In this case, it will be understood from what precedes that a sustained yield will be assured if the possibility is calculated from the volume of the full-aged crops divided by one-third the number of years in the exploitable age. Since no allowance is made for the increment, it may safely be assumed that the yield so calculated need never be reduced in subsequent revisions of the possibility.

But the normal proportion of 5 to 3, between the volumes of full-aged and medium-aged crops, will not often be found to exist in India. There may be—

(*a*) excess of full-aged crops, or
(*b*) excess of middle-aged crops.

Where the volume of the old crops is abnormally large the condition of the smallest trees in the full-aged crops should be examined with a view to ascertaining whether it will be practicable to keep these stems standing till the medium-aged crops are exploited. If it appears that this can be done, a portion of the full-aged crops should be placed in the group of medium-aged crops. In other words, it should be sought to make good the deficiency of medium-aged crops by adding a given amount of wood from the full-aged crops.

Where the volume of medium-aged crops is in excess, it should be ascertained whether the largest trees in the medium-aged crops cannot be exploited during the period assigned for the felling of the full-aged crops, and it should then be sought to supply the deficiency of full-aged crops by transfers.

When the possibility has been determined, the length of the felling rotation, during which the whole forest will be gone over by the selection fellings, should be fixed. In the present case it should, however, be one-third or a sub-multiple of one-third of the exploitable age.

To this method of determining the possibility of selection-worked high forests the following advantages are ascribed :—

(a) It allows the possibility to be fixed according to the state of the crops on the occasion of each revision of the working-plan.

(b) It tends to introduce the normal proportion between the different age-classes or size-classes of trees.

(c) It helps to secure a sustained yield, so far as the composition of the forest renders this possible.

Fellings limited by proportionate volume.—This method of determining selection fellings by volume of material, known in France as the *système Masson*, is based on the assumption that as, in a high forest worked by the regular method, a definite proportion of the crop on the ground is always felled, so, in a selection-worked forest by invariably limiting the fellings to a certain percentage of the volume of the crop the fellings may be kept within the possibility.

Applying this theorem to a fairly homogeneous fir forest, containing a wooded area of 1,600 acres in which the average annual production or capability of the soil has been ascertained, in the manner explained at page 89, to be 63 cubic feet of wood per acre a year, and assuming that the forest is worked by the regular method on a rotation of 150 years, each acre would produce during the rotation 150 × 63 = 9,450 cubic feet, and the volume of wood at any time during this rotation would, on the whole area, be one-half of 1,600 × 9,450 or 7,560,000 cubic feet. Now, the area of the forest being 1,600 acres and the number of years in the rotation 150 years, the size of the annual *coupes* (the forest being worked by the regular method) would be 1,600 ÷ 150 = 10·66 or 10⅔ acres, and the volume of material felled each year would be 10⅔ × 9,450 cubic feet = 100,800 cubic feet, or 1·33 per cent. of the total material in the whole forest.

Similarly, were the rotation reduced to 100 years, the volume of wood on the ground would be (63 × 100 = 1,600) ÷ 2 = 5,040,000 cubic feet ; while the material removed at each felling, made on an area of 1,600 ÷ 100 = 16 acres, would still be 16 × 63 × 100 = 100,800 cubic feet. In this case, 2 per cent. of the material on the ground would be felled.

Therefore, in a high forest worked by the regular method, there is a fixed proportion, depending on the length of the rotation, between the volume of the material felled each year and the total volume of material or the wood capital of the forest. This proportion, following a general law, varies inversely as the length of the rotation ; that is to say, it is higher as the rotation is shorter and *vice versâ*.

Let us suppose that it is wished to ascertain by the volumetric method the possibility of the fir forest of 1,600 acres, excluding blanks, for which it has been ascertained that the productive power of the soil is 63 cubic feet of wood a year. Let us also assume that the exploitable size of the trees is 2 feet in diameter, and that it requires on an average 150 years for the trees to attain that dimension.

Under these conditions the material on the ground being, as stated, (150 × 63 × 1,600) ÷ 2 = 7,560,000 cubic feet, and the size of the annual coupes being 1,600 ÷ 150 = 10⅔ acres, there would be felled each year 10⅔ × 63 × 150 = 100,800 cubic feet of wood, or 1·33 per cent. of the total material on the ground. On the assumption that were the forest worked by the selection method, an equivalent yield could be obtained, it would be possible, without exceeding the capability, to fell 1·33 of the material on the entire area, or 13·3 per cent. of the material on one-tenth of the area, every ten years. We may suppose that number of years to be the interval between the successive fellings on the same area and the forest to be divided into ten coupes, or we may take as many times 1·33 per cent. of the material as there are coupes or years in the interval between the successive fellings. Now the wood capital will be far from being exactly and evenly distributed over the forest; some of the compartments will be rich in material and some poor. Each year there would, however, be exploited successively in each coupe in its turn 13·3 per cent. of the material on the ground, without ever exceeding this figure, but so as to fell exactly 100,800 cubic feet of wood. If the 13·3 per cent. of the material in a coupe exceeded the allotted total volume of 100,800 cubic feet, the surplus would be left for the following year; if less the volume prescribed would be made good from the next coupe. In this manner what was felled in a compartment would always bear a fixed ratio to what existed in that compartment; from the rich compartment more would be taken, from the poor less; while the outturn could remain the same from year to year.

Neat and exact as this volumetric method of calculating the capability may appear in theory, there are several drawbacks to its practical application, especially in Indian forests. It is certainly superior to other methods, in that it secures an absolutely equal outturn from year to year; but this in India is often a matter of comparatively small importance. On the other hand, it not alone requires nice calculations based on exact enumeration and measurement of the whole standing crop—and this difficult and expensive undertaking must be repeated periodically—but it also necessitates fellings entailing most careful supervision being made outside the coupe of the year.

Altogether the method is in India inferior to that of prescribing the capability by number of trees. Any mistake in estimating the production of the soil—always a difficult calculation to make practically, however simple it may appear in theory—would, under the volumetric method, have a very injurious effect on the crop. This, as already explained, cannot occur where the number of trees has been fixed without limit as to size; for if the possibility were over-estimated, smaller trees, and consequently a less volume of material than the calculated capability of the forest, would be felled.

It will be observed that the last two methods of determining the capability are based on the assumption that all sizes of trees are saleable, and that the forests to which the methods are applied are in the normal condition of selection-worked crops; that is to say, containing a complete series of trees of all ages scattered irregularly over the entire area. There are, however, comparatively few forests in India in which these two conditions are found to co-exist. For where the demand is good, the forests have for the most part been so overfelled and mal-treated that they contain little or no sound mature stock; and where the crops have not been mal-treated, there is generally no demand for any but the principal species.

Length of the felling rotation in selection-worked forests.— It will have been seen that the application of the possibility involves the determination of the felling rotation; and in fact, the first step towards introducing order in the working consists in restricting the locality to be worked over each year or period to a definite area. The object of this is to keep the extent of each separate felling within manageable limits; to conduce to effective supervision of the work; and to render the extraction of produce easier and less expensive. Theoretically, every acre of the whole area ought to be worked each year by the removal of the exploitable trees. But this would not be practicable; as the trees to be felled would be scattered over so large an area that it would not be profitable to remove them. It is, therefore, necessary to decide what time may be allowed to elapse between successive fellings in the same place, and, having determined this, to divide the forest into a corresponding number of coupes, one of which will be exploited each year or period.

Where the possibility is calculated by the usual Indian methods and prescribed by number of trees, it may be found convenient to make the length of the rotation equal to, or a sub-multiple of, the average number of years required for trees of the lowest girth of the second class to attain the lowest exploitable size. In other cases it may be found that the felling rotation should, in order that the working-plan may be as simple in construction as possible, be a sub-multiple of the *exploitable age*. This can easily be done by slightly enhancing or reducing the calculated exploitable age. There is nothing wrong in such an adjustment; for it is ridiculous to suppose that the calculated rate of growth is so accurate that a variation of a few years either way will make any material difference.

Within limits, it is a matter of comparatively small importance from a sylvicultural point of view at what intervals of time fellings are repeated in the same area. With a short as compared to a longer interval, the area gone over at each felling is larger, and the quantity of produce removed is less per unit of area; while the crop has a shorter time in which to recover from any injurious effects attributable to the operation of felling. Practical considerations, as regards transport, supervision, etc., should for the most part, determine the question. The interval should be long enough to prevent injury to the crop as a whole from the too frequent repetition of fellings in the same area, and should be sufficiently short to enable dead or dying trees to be removed in good time. In some Indian working-plans as long an interval as 30 years has been adopted. Usually, however, not more than 10 or 15 years is taken ; and, where the demand is intense, even a shorter interval may be adopted with advantage.

<small>This determination of the rotation of the fellings and the corresponding division of the forest into coupes is essential to methodical working. In some cases, in India, the application of the selection method has merely consisted in the prescription of certain cultural rules without any annual coupes. Such plans miss the point : they do not serve what ought to be their chief purpose, namely, the introduction of order into the working.</small>

Selection fellings.—The fellings should always be prescribed for one or more complete felling rotations as in that period they should pass over the whole area of the working-circle.

The area to be taken in hand is determined, as has been described for simple coppice, by dividing the working-circle into as many areas of equal forest resources as there are years or periods of years in the felling rotation. There is, however, this difference that these areas are often for instance in a mixed crop for which one species out of many is exploited—exceedingly large ; so that instead of each block containing a number of coupes, each coupe may, in some cases, include a number of blocks.

The order of the fellings is of less importance in the selection method of treatment than in any other. It will, however, generally be convenient—and there will rarely be reasons for adopting another course—that the fellings should follow one another in regular order on the ground.

Whatever system be adopted in order to calculate the possibility, the fellings should be prescribed in a simple manner and according to the principles of the selection method. The general rules for the application of those principles may, where necessary, be supplemented by cultural prescriptions for each compartment.

Restoration of an incomplete crop.—It is of frequent occurrence that the crop to be dealt with is so abnormal that the selection method, pure and simple, cannot be directly applied to it. In such cases the crop should be subjected to fellings regulated by cultural rules, in the manner already explained, until the forest capital has attained the necessary composition.

The length of the period in which the restoration may in practice be made can generally be determined from an inspection of the crop. At most, an enumeration of the sound stock in the forest, and a determination of the rate of growth of the principal species, are required. In cases of restoration the fellings of mature sound trees would be made with great moderation, and be strictly limited to the removal of such stems as were over-mature or approaching decay. Otherwise, if all trees were removed as they became exploitable, the object sought would be defeated because the wood capital on the ground would not increase. It is of especial importance to note this, because by the lapse of time there would be more sound trees to fell and fewer unsound. There is then temptation to fell the sound trees and to neglect the unsound. This should be carefully guarded against if the restoration fellings are to have the desired effect.

It not infrequently happens in India that the demand for timber is so uncertain and so unequal that if fellings in fixed areas were prescribed in advance the material cut could not be sold. It has been sometimes sought to get over this difficulty by prescribing the areas to be taken in hand each year, leaving the executive officers to make the felling heavy or light according to the demand at the time. This, however expedient it may be from a revenue point of view, is nevertheless sylviculturally a faulty system. Theoretically the state of the crop determines the nature of the felling. The object of restoration fellings during a provisional rotation is to assist in bringing the crop to its normal condition. To regulate the intensity of the fellings by the demand might have just the contrary effect, and might defeat the chief object which the plan should have in view. In practice, however, it would often be mischievous to fell trees which might be profitably disposed of hereafter. The conditions of each case must be weighed ; and it may sometimes be advisable to confine the restoration or improvement fellings to the trees least capable of improvement, having other stems, which in strictness ought to be removed for possible sale in the future.

The transformation fellings should be prescribed by area in the manner already explained for fellings regulated by the cultural method.

METHOD OF SUCCESSIVE REGENERATION FELLINGS.

General description.—The essential cultural feature of thi method consists in the gradual removal, by successive fellings over definite and limited areas, of the crop to be regenerated. While this regeneration by successive fellings is proceeding, the crops which have been or have still to be regenerated are subjected to cleanings and thinnings. The principal or regeneration fellings are prescribed by volume, that is to say, the possibility in cubic feet having been ascertained, as many trees as are found to yield this volume are removed wherever, in view of the state of the crop, this is required for sylvicultural reasons.

Possibility : volumetric method.—Formerly the volume of material to be removed annually was determined by the volumetric method, which consists in calculating the total volume of wood produced in an interval of time equal to the exploitable age, and by dividing this total by the number of years in that age. The method is still sometimes employed.

The manner in which the calculation is effected may best be explained by an example. Suppose a forest of 1,600 acres, which it is wished to exploit at 120 years, divided into compartments or cultural sub-divisions each containing a crop of an uniform age. The exploitable age would be divided into, say, four periods of 30 years each, and the area, *provisionally*, into four corresponding blocks, the first containing the crops above 90 years old, the second those of 60 to 90 years, the third those of 30 to 60 years, and the fourth those below 30 years.
Thus :—

1st Block.

Compartments	1 and 8	88 acres.	130 years' old.
,,	2, 3, 5	184 ,,	120 ,, ,,
,,	4 and 9	68 ,,	110 ,, ,,
,,	6 and 11	84 ,,	100 ,, ,,
,,	7	48 ,,	90 ,, ,,

And so on for the other blocks. This preparatory classification having been effected, the volume of material in each compartment is calculated, and to this volume is added the estimated future increment up to the time when the compartment will be regenerated, that is to say, 15 years' growth (the mean of the 30 years during which all the compartments would be regenerated, some at the beginning, some at the end of the period) for the compartments forming the first block, 45 years' growth for the compartments forming the second block, 75 for those forming the third block, and 105 for those in the fourth block, thus :—

1st block,	actual volume	1,386,000 c. ft.	estimated growth 15 years,	210,000 c. ft.	total	1,596,000 c.ft.
2nd	,,	,,	393,000 ,,	,,	,,	45 ,,	720,000 ,, ,,	1,113,000 ,,
3rd	,,	,,	435,000 ,,	,,	,,	75 ,,	1,245,000 ,, ,,	1,680,000 ,,
4th	,,	,,	262,000 ,,	,,	,,	105 ,,	1,419,000 ,, ,,	1,681,000 ,,

Total ... 6,070,000 c.ft.

Consequently, the total volume to be felled during the first period of 30 years, after which the calculations would be revised, is 6,070,000÷4 = 1,517,500 cubic feet, and each year 1,517,500÷30 = 50,583 cubic feet. The permanent plan would then be drawn up by allotting, to the first block, to be regenerated during the period of 30 years, a sufficient number of the compartments containing the oldest crops to furnish, including future increment, 1,517,500 cubic feet of material. As the compartments *temporarily* included in the first block will furnish 1,596,000 cubic feet, certain crops should now be allotted to the second block instead. There would then be felled each year during the period, in the compartments forming the revised first block, 50,583 cubic feet of material. During the first period, thinnings would be carried out in the compartments of the other blocks until the end of the 30 years, when the compartments in the second block would be taken in hand and dealt with in the same way.

This method of calculating the yield is now, owing to the difficulty of accurately estimating for long periods the future growth, seldom used except in certain parts of Germany, where, however, numerous improvements have been introduced in the manner of verifying the possibility and in graduating the age-classes. These modifications are, however, too complicated for adoption in this country, where indeed the method is not generally suitable in present circumstances, owing to its complicated nature, its uncertain results, the frequent revisions necessary, and above all the irregular nature of most of the crops dealt with.

Possibility: mixed method.—The volumetric method has been generally replaced by a simplified form, known as the mixed method, the essential feature of which is the two-fold and simultaneous division of the exploitable age and the area into corresponding portions. In other words, instead of calculating the future growth of *all* the crops up to the time of felling, the total area is divided into a number of equi-productive blocks, and the exploitable age into a similar number of periods during which each block is in its turn to be regenerated. The yield of that block which is about to be taken in hand for regeneration is then calculated.

Thus, in the example already taken, instead of calculating the future growth of the crops in all the blocks, and then making transfers from one block to another so as to secure an equal yield throughout, the forest area is partitioned off into four blocks, each about 400 acres in extent or each of equi-productive resources.

Thus, block I would contain the mature crops.
" II " " large pole crops.
" III " " younger pole crops.
" IV " " young growth.

In this partition the size of each block, where the elements of production were not uniform, would be enlarged or reduced, as the case might be, so that all might eventually yield about the same quantity of material. The yield of the first block only would then be calculated. This would be done, as in the previous example, by adding to the material in the ground the estimated growth during half the length of the period—in this case close on 15 years—and by dividing the total so obtained by 30, the number of years in the period. The future growth is obviously taken as the average growth of the whole for *half* the period only; because some of the crops would be exploited at the beginning and some at the end of the period.

The trees in the block under regeneration would, as already stated, be generally removed gradually in several successive fellings. The first, the preparatory or seed felling, would be made with the object of opening out the canopy so as to obtain young seedlings sufficiently lighted for proper growth. The extent to which this felling should interrupt the cover would depend on the species constituting the seedlings, the nature of the soil, and the climatic conditions. Where the cover was only slightly interrupted, the felling would be said to be close; otherwise it would be an open seed felling. When the crop of seedlings became fairly complete, the young plants, having reached a certain age, would require more light. This would be effected by one or more secondary fellings. A final felling, to remove the trees not cut in the secondary fellings, would be made when the soil was fairly covered with young growth in the thicket stage and had nothing to fear from complete exposure. During the interval between the first and final fellings the young seedlings would be carefully fostered by cleanings. These would be made principally with the object of removing or retarding the upward growth of the more vigourous but valueless species, which nearly always invade the ground and threaten the existence of the more important kinds. The cleanings would be followed by thinnings, to be continued until the crop was almost mature. In the blocks containing medium-aged crops not yet under regeneration, thinnings would also be carried out, and where, as often happens, overmature trees were scattered through the crop, these would be removed. The thinnings would be made with a view to preparing the crop for regeneration during the period assigned by the plan for that purpose, and generally with the object of favourising the better trees by the timely removal of less promising stems. Thus, in the block to be regenerated in the following period, only dead and decaying trees would be removed, while in those blocks which were not to be regenerated for a considerable time, all trees over a certain size or age, up to the limits of the capability of the forest, might be felled, provided their removal was sylviculturally desirable in the interests of the trees of the future.

This is an outline of the manner in which the method of successive fellings is organised when applied to a forest in which crops of all the different age-classes exist in the requisite proportions. It rarely happens, even in Europe, where the forests contain distinct age-classes, that the method can be directly applied without some deviation from the ideally correct or normal treatment. Generally speaking, in order to form the periodic blocks of compact shape and fairly equal area, it is necessary to include in them crops which must be regenerated *out of their turn* so to speak. An example of such a proceeding is here given; but obviously the variations, which circumstances may render necessary in applying the method, are endless.

Example of the general working scheme for a forest of 1,786 acres, treated as high forest by the method of successive fellings and thinnings under a rotation of 150 years divided into five periods of 30 years each.

No.	Area to be taken in hand.			Period in which to be regenerated.	Remarks.
	Blocks.	Compartments included.	Acres.		
I	Deota	A 1	54	1880-1909	The compartment C1 has already been regenerated and contains only young seedling growth.
		B 1	87		
		C 1	214		
II	Chansil	A 2	176	1910-1939	
		B 2	149		
	Noranu	C 2	105		
		D 2	33		
III	Datmir	A 3	153	1940-1969	
		B 3	117		
		C 3	80		
IV	Bamen	A 4	189	1970-1999	
		B 4	68		
		C 4	100		C5, though placed in the 5th block, contains mature high forest, which it will be necessary to regenerate in the first period, and to exploit a second time in the fifth period.
V	Sahlra	A 5	205	2000-2029	
		B 5	112		
		C 5	44		

Working scheme for the first period of 30 years, 1880 to 1909.

Periodic blocks.	Compartments.	Areas.	Ages of the crops in 1880.	Nature of fellings to be made.	REMARKS.
I V	A 1 B 1 C 5	54 87 44 ——— 185	140 130 150	*I.—Fellings by volume.* Regeneration fellings by volume, 35,893 c. ft. a year during the first ten years until the first revision of the possibility.	The fellings by volume comprise seed or preparation fellings, secondary and final fellings. The actual volume of wood in compartments to be subjected to these fellings is 823,300 c. ft. Adding the estimated growth during 15 years (one-half of the period, as some fellings will be made at the beginning and some at the end of the period) at the rate of 2 per cent. a year, or 248,490 c. ft., the total volume is 1,076,790 c. ft. Therefore, during each year of the period there may be felled 1,076,790 ÷ 30 = 35,893 c. ft.
II	A 2 B 2 C 2 D 2	176 149 105 33 ——— 463	125 110 85 90	*II.—Area fellings.* Final thinnings by one-fifteenth of the area (31 acres a year) repeated twice during the period.	
III IV	A 3 B 3 C 3 A 4 B 4 C 4	153 117 80 189 68 100 ——— 707	80 75 70 50 65 40	Thinnings by one-tenth of the area (71 acres a year,) repeated three times during the period.	
V I	A 5 B 5 C 1	205 112 214 ——— 531	6 15 16	Cleanings according to cultural requirements.	The possibility will be revised at the end of ten years.

The fellings.—The fellings in this method are prescribed for one period only. The length of this period, which is a sub-multiple of the exploitable age, should be long enough for the seedlings of the principal species to establish themselves completely.

Except in the modified method subsequently mentioned, the area to be taken in hand each year or sub-period is not prescribed for the *principal fellings*. These are, it will be seen, prescribed by volume and may be made wherever required in any part of the periodic block to be regenerated during the first period. Selection fellings and thinnings are prescribed by area in the usual manner.

The order to be followed and the nature of the fellings to be made require no further explanation.

Modifications of the mixed method.—Several modifications have been recently introduced in France into the application of the method of successive fellings. The principal of these consists in the establishment from the first of annual coupes, instead of leaving the position of the coupes to be fixed from year to year by the controlling officers according to cultural requirements. The whole area is divided off into as many permanent annual coupes as there are years in the exploitable age. The coupes are made fairly equi-productive, by deduction of blanks and by assessing the fertility of the great natural divisions of the forest in the manner already explained for coppice. The coupes being formed, the preparatory or seed fellings are then prescribed by *area*, one of the coupes being taken in hand each year for this purpose. The same system is followed with regard to the thinnings and cleanings, which are made by area in regular succession in one of the annual coupes. But the basis of area has to be somewhat departed from as regards the secondary and final fellings which depend on the state of reproduction and on the development of the seedling crops. These fellings must be regulated by volume as in the old method, and for this purpose the coupes in the block under regeneration are grouped together, and the volume to be felled each year in this block is then calculated in the manner already explained.

Application of the method to irregular crops.—None of our Indian forests could be subjected directly, in their present condition, to the method of successive fellings, as they do not as yet contain definite groups of age-classes. Should it be decided to apply the method to any of these forests, it will, therefore, first be necessary to *transform* the crop into one of the type required, that is to say, one containing a regular series of age-classes. For this purpose the first thing to be done is to lay down the general frame-work of the plan according to the method of successive fellings which it is hoped eventually to apply. The general working scheme would be framed and the circles and periodic blocks or coupes corresponding to this frame-work would be laid out. Until this is done, and the place which each portion of the forest will occupy in the final scheme is known, the cultural rules to be applied to each crop cannot be prescribed. The general working scheme gives the order in which the various portions of the working-circle are to be taken up for transformation during each successive period.

The formation of the working-circles is not a matter of difficulty. A correct gradation of ages is by hypothesis unobtainable. The duration of the provisional plan for the transformation of the forest would be equal to the exploitable age of the forest which it is sought to constitute, and would be determined in the same way, that is to say, it would depend on the size of the exploitable trees and on the rates of growth. The periodic blocks would consist of a few great natural divisions of the forest, chosen so as to furnish more or less equal material during corresponding sub-multiple periods of the exploitable age. Indeed, the general scheme is usually very obvious and well defined. When it has been thus drawn up, two classes of operations will be required, namely, (1) transformation fellings in those portions of the forest which are to be first regenerated, and (2) modified selection fellings elsewhere.

The transformation fellings do not usually differ very materially, at least in principle, from ordinary successive regeneration fellings; but their execution will depend more markedly on the nature of the crop, and it will be necessary to change the manner of applying them from place to place according to the irregularities met with. Where reproduction is abundant, they will simply remove the mature trees in whatever way is required, so as to leave room for the young growth to develop. Where there is little or no young growth already established it will be sought to obtain reproduction in the ordinary way by successive fellings, preparatory or seed, secondary and final. The selection fellings to be carried out in the blocks to be regenerated later on offer no difficulty. The trees to be removed will be such as cannot maintain themselves in a sound state until the crop in which they grow reaches its turn for regeneration. Knowing the length of the rotation and the rates of growth, the question is easily determinable. Thus, where the exploitable size of the trees is 24 inches in diameter, with a rotation of 120 years, and a rate of growth of 10 years for each inch of radial increase, all trees above 12 inches in diameter might be felled in blocks or coupes in which regeneration fellings will not be made during the following 60 years; while in those blocks which will come under regeneration in, say, 30 years, only absolutely unsound trees, or at any rate trees well over 18 inches in diameter, would be felled. The selection fellings would in other respects be conducted as in ordinary selection worked forests regulated by area according to the cultural method.

The transformation fellings would be regulated by volume in the manner already explained. The selection fellings would, on the contrary, be applied by area according to cultural rules, with or without a limit as to the number of trees to be felled. The general working scheme will be drawn up absolutely as already explained.

THE GROUP METHOD.

General description.—This is merely a modification of the method of successive fellings. All the calculations connected with the manner of prescribing the possibility, etc., are identical in the two methods, and the gradual exposure of the young seedlings is secured by means of two or three successive regeneration fellings. These successive fellings are, however, in the group method all made simultaneously by groups, that is to say, wherever a patch of seedlings is already established it is at once exposed in greater or less degree to the climatic influences. Where seedlings do not exist, the cover is opened out with a view to their establishment. The method in fact amounts to this that the volume of material to be felled annually in the block under regeneration having been determined, this volume is felled by patches or groups wherever a group of established seedlings exists or is desired. The patch is then enlarged, or new patches are opened out from time to time as the area covered with established seedlings extends. The method is applied in exactly the same way as the method of successive fellings, and comprises similar clearings, thinnings, and selection fellings.

PASTORAL TREATMENT.

Areas to which applicable.—There are many tree-clad lands in India under the control of the Forest Department which can be profitably utilised only by the pasturing of cattle, and which, in the interests of the community, should be devoted to the production of fodder. Such areas require, however, to be placed under forest treatment, with a view to the preservation of the trees whose roots penetrating to deeper and fresher soil maintain active vegetation in the dry season, thus enabling the overhead cover to protect and conserve the shallow-rooted grasses which would otherwise inevitably succumb.

As cases in which the pastoral method may have to be organised, there may be cited those tracts in some of the drier parts of India where the land, except when irrigated or adjoining perennial rivers, is unculturable. The rainfall is insufficient, and the water level is too far below the surface for any system of well-irrigation to be profitable. Cultivation is, therefore, confined to the banks of rivers; but, even there, gives a poor return. Large numbers of people, however, frequently manage to subsist in these areas owing to the addition to their income and food-supply derived from their herds and flocks. These animals obtain food from the trees and bushes dotted about in the uninhabited and unculturable tracts away from the rivers, and which even in years of drought, owing to the depth to which their roots penetrate in the soil, yield some leaves which can be used as fodder. In such cases the grazing lands are the very life of the people, but should the trees on them disappear they would become almost absolutely barren. It is, therefore, necessary in the interests of the people to subject the lands to forest management, and so restrict grazing so far as may be necessary to ensure the maintenance of the trees. It should not be sought, however, to produce timber except such as may be yielded by some method of treatment that will not interfere with grazing. These lands in many cases only produce annually some two or three cubic feet of wood worth not half as many annas. The grazing fees, if the number of cattle were limited to what the lands could support, might well bring in as much. Apart, however, from considerations of State revenue, if the number of people benefited by the grazing, as compared with the number benefited by wood culture, is taken into account, the result is much in favour of pastoral treatment.

Application of the pastoral method.—The treatment of Indian forests, primarily with regard to the fodder they produce, has been perhaps insufficiently studied in view of the enormous importance of such areas in the economy of the country. It may be accepted that the greatest quantity of fodder will be obtained by absolutely closing the area to cattle grazing and by allowing grass-cutting only. For it is indisputable that animals, while grazing, trample and destroy more grass than they eat, damage young seedlings and trees, and harden the soil. But the exclusion of cattle and the enforcement of grass-cutting is not often feasible, owing to the distance to which the fodder would have to be carried and to the small value of the animals. There are many areas which it would be impossible to utilize except by cattle grazing. In such cases it will generally be possible to sufficiently restrict injury to the trees and to ensure their reproduction and perpetual maintenance by—

(1) limiting the number, or the kinds and classes, of animals grazed; or

(2) limiting the period during which grazing is allowed; or

(3) combining (1) and (2).

The first method, where it can be effectively carried out, is the best—the best not merely in the interests of the vegetaion but also of the animals, which are thereby well nourished. But it requires, as compared with the second method, a larger and better supervised establishment; and the small fees usually realised for grazing will rarely cover the expenditure involved. In practice, therefore, the second method, although much less effective, is often preferable. Sometimes both methods may be enforced simultaneously, as in certain of the Ajmere forests.

It may be remarked in this connection that one of the chief causes of the inferiority of Indian cattle is the smallness of the fees charged for grazing and the loose way in which these fees are collected. The Indian cattle-owner does not limit the number of his stock according to the grazing area at his disposal. He usually lets nature perform this office, and keeps as many ill-fed or half-starved animals as can manage to escape death. This would not be the case were the fees charged sufficiently high and were the number of animals allowed to graze per acre strictly limited. It is a common practice for Government to sell the grazing right by auction, or to rent it to a contractor on the condition that the fee charged *per capita* shall be limited to a certain maximum. The contractor's interest then is to graze as many animals as possible on the land. The appalling number of cattle which are killed off every dry year in the more arid parts of India has led to many proposals for increasing the grazing areas and for throwing open the Government forests to grazing. Such measures merely tend to increase the number of miserable animals, which exist without profit and even to the detriment of the owners until the next season of drought does its work. The only remedy appears to lie in limiting the intensity of the grazing in the Government estates to what the land can support, and in charging such fees as will make it worth the while of cattle owners to keep only animals that can bring them in a profit.

Although the areas subjected to the pastoral method of treatment should be worked with a view to the production of fodder, it by no means follows that none of the wood produced can be utilised. Apart from the fodder, a supply of small timber can be obtained by subjecting the trees to branch coppice fellings for instance, or, where the number of cattle is limited, over-mature trees can be removed by the selection method. It should not, however, be overlooked that the treatment is primarily intended to favour the production of fodder not of wood, and that if restrictions are introduced with the object of preserving the trees it is because these trees are chiefly useful for the fodder they furnish, either directly in the shape of leaves and twigs or by the shelter they afford to the soil. In most cases it will be found that the direct profit from grazing is greater than that which could be obtained under the strictest system of closure for timber growing.

SUPPLEMENTARY PROVISIONS OF A WORKING SCHEME.

Subjects to be dealt with.—The application of the method of treatment adopted may involve thinnings, cleanings or even selection or other fellings being carried on in various parts of the working-circle, blanks or other areas being re-stocked artificially, or valuable species being introduced into the crop. It will also generally be necessary to regulate the grazing of cattle and the removal of produce by right-holders, or of dead and fallen wood; and, as a rule, to provide generally for certain improvements, such as the construction of roads or buildings, for fire protection, etc.

Improvement fellings.—These, if any are to be made, will have been indicated with the method of treatment. In all cases they will be prescribed by area.

Sowings and plantings.—The importance of artificial reproduction in the mode of working will be indicated in the method of treatment on which the necessity or otherwise of such operations depends; because regeneration may be entirely obtained by artificial re-stocking, or the operations may be confined merely to the re-filling of a few blanks. The introduction of superior species into a crop will generally be carried out by planting.

Regulation of grazing.—Forest grazing in areas which have been permanently set aside for the production of wood should not be confused with the pastoral method of treatment already described. The two classes of forests are absolutely distinct as regards the purpose for which they are managed. It may be assumed that, as regards the former areas, it is in the interests of the community at large, if not of the population immediately surrounding the forests, that the forest should produce timber, large or small. It may not be necessary or expedient to exclude grazing altogether from such areas ; but there should be no room for doubt as to whether grazing or the production of wood is to give way.

Many of the troubles of the Forest Department are due to a proper distinction not being made between lands which, in the interests of the country, should be managed with a view to the production of wood and those which should be devoted to the production of fodder. Local Governments hesitate to place lands under forest management, because experience has shown that lands so placed are liable to be turned into close preserves for the production of timber—the supply of which is often already too abundant for local consumption—and that thus the grazing requirements of the people are overlooked. The confusion between the two classes of areas has not infrequently ended in an unhappy compromise under which neither one class nor the other can be properly managed. For it is as unwise to unnecessarily restrict grazing in fodder reserves as it is to admit unrestricted grazing into timber forests. In preparing a working-plan the two classes would necessarily be separated ; but, unfortunately, many of the forests have already been burdened with permanent rights of grazing which seriously interfere with their utility and which would never have been imposed on them had a wiser and more enlightened policy been generally followed with regard to the recognition of the natural grazing grounds.

Grazing, whatever may be the circumstances which render it necessary to admit cattle into the forest, should be regulated and organized in the working-plan that is prepared for the area burdened. It is, therefore, desirable to consider the restrictions that may have to be imposed and how they should be regulated. In the first place an endeavour should be made—so far as this is possible in view of admitted rights and, in some cases, local customs as strong as rights—to apply the following principles :—

(1) A distinction should be made between the grazing of sheep and goats, and of cattle, horses or donkeys which do much less harm ; and goats and sheep should not be allowed into any area placed under regular forest treatment.

(2) Every herd or flock permitted to graze should be placed in the charge of a shepherd, who would be responsible for any infringement of the grazing regulations and for any injury done by the animals in his charge.

(3) All portions of the forest undergoing reproduction or containing very young growth should be entirely closed to grazing.

(4) In the lower hills, the upper strip of forest bordering on the pasture land should, in most cases, be closed to all grazing.

(5) The number of animals allowed to graze should be proportionate to the area concerned and to the quantity of fodder available.

(6) No animal should be allowed into the forest before the season at which the young crop of grass of the year appears.

As regards the exclusion of goats and sheep, it is not meant that these animals must necessarily be kept out of all wooded areas. This is not always possible or advisable : what is meant is that wooded areas within which browsing animals are allowed cannot be subjected to any regular method of forest treatment. Money spent on works of forest improvement in such areas will generally be wasted, and the best thing will be to abandon the area in question as 'forest.'

Trees are not safe against injury from cattle, sheep, etc., until they have attained a sufficiently considerable size for their young shoots to be out of reach. Even then damage, especially as regards reproduction, is caused by the trampling and hardening of the surface-soil. Not merely forest therefore which have been felled, but also those which are about to be felled, should generally be closed to grazing in order to enable the soil to attain the condition suitable for the germination of seed. It thus follows that it is only in forests which are exploited at a considerable age that grazing to any great extent is possible. Coppice compartments exploited at a low age, can at the best be only opened for a very few years at a time. Forests worked by the selection method ought, strictly speaking, to be always closed ; because reproduction is going on all over the area. Where total closure is impossible, means must be taken in some cases to exclude grazing from certain areas which are about to be felled. Closure against grazing can, in the absence of fencing, only be effectively carried out if the portion closed has good natural boundaries, such as streams or deep ravines. Hence the importance, for one reason, of closing as blocks *natural* sub-divisions of a forest. Blanks requiring to be restocked are often a serious source of difficulty in areas open to grazing. The blanks cannot be re-stocked if grazing, which may indeed have caused the blanks, be not excluded. It will often be necessary in a working-plan to arrange for fencing such areas, or for separating by a fence those portions of a forest which are open from those which are closed.

It should not be overlooked that restrictive measures should be introduced gradually. Their introduction is a question of time and especially of tact. But the drier and hotter the climate, the greater is usually the necessity for closure.

In addition to these measures efforts should be made to develop the value of the forest produce by means of roads.

People respect property of real value. It has been said that the chief reason why the forests in the mountains of Auvergne and in the Alps were destroyed by excessive grazing was because the wood was without value in such out-of-the-way places; while forests round Paris have been preserved because there has always been a good market for their products.

It may not here be out of place to indicate briefly the legal restrictions imposed in Europe where the subject of grazing has long received attention. Taking France as an example, the regulations regarding grazing contained in the Forest Code of 1827 date from very distant times. In 1541, Francis I, revived the decrees on this subject, previously in existence, and these were again sanctioned in 1660 in the celebrated Forest Ordinance of Colbert. The present Forest regulation on the subject is merely a repetition of these ancient laws. The most important provision in the old laws consisted in the power to close to all grazing for a definite period certain portions of a forest or even the entire area, if such a measure appeared necessary for the safety of the forest. Under the existing law, instead of closing certain specified areas, the areas open to grazing, in which the right-holders are allowed to graze, are fixed each year by the Forest Officers (Articles 67 and 69), as also (Article 71) are the roads and paths (by which the animals grazed are allowed to pass through the closed portion of the Forest), the number and kind of animals which the right-holders are to be permitted to graze, and the time during which they can be grazed. *The ntroduction of goats is absolutely prohibited notwithstanding any title to the contrary.*—Article 78. The same restriction is placed on sheep; but, in cases where there are no other means of providing for their support, the Government may permit cultivators, living on the borders of the forest, to pasture their animals in the portions open to cattle grazing.

As regards animals other than sheep and goats, the Forest Officers fix the number that may be grazed each year, and the period during which grazing may take place. This number is necessarily proportionate to the area thrown open (Articles 65, 66 and 68). Right-holders are allowed to graze animals required for their *bonâ fide* domestic purposes, but not animals kept for trade or speculation (Article 70). In order that the Forest Guards and Inspectors may be able to recognize them, all animals entitled to graze in the forests are branded (Articles 73 and 74). The animals belonging to one commune or group of right-holders are all grazed together under one shepherd (Article 72) appointed for that purpose. In this the modern law has followed ancient regulations, as it was found that with a number of animals scattered through the forest supervision became impossible. Separate groups of right-holders are not however, allowed to unite their cattle under one shepherd, as too great a number of animals grazing together injure the soil.

The shepherds appointed are directly responsible for breaches of the regulations, or for injuries done by the animals in their charge; and, if fined, the commune appointing them are responsible for the payment of the fines (former Article 72).

It is provided by Article 119 of the general rules made under the Law that every year the local Forest Officers shall, having due regard to the nature, age and situation of the trees, report, in a formal written proceedings, the condition of the blocks of forest under the *régime forestier* which can be made over for grazing. They are to indicate the number of animals that can be admitted to these blocks, and the dates on which the exercise of the rights of user may commence and must end. The proposals of the Forest Officers are submitted for the approval of the Conservator before the 1st February in each year.

There are very severe penalties for a breach of any of these grazing rules. Right-holders introducing goats or sheep are subjected to a double fine. Right-holders who introduce more animals than they are entitled to, or who graze in closed portions of the forest, are treated as if they had no rights, and are subject to the same penalties as if they were outsiders.

Regulation of rights in wood.—Where there is merely a right to a certain *quantity* of produce from any part of a forest there is no difficulty, as the produce would come from the prescribed fellings for the year in the coupe set apart for the purpose. But in India it rarely happens that the right is in this form : the right is usually a servitude over a comparatively small area. In such cases it is not so easy to arrange for the supply to a number of scattered villages of the given quantity from the coupe of the year. Where the demand is inconsiderable, it may often be conveniently met by fellings of scattered trees independent of the main fellings. The produce delivered to right-holders must either be deducted from the available crop, in calculating the possible yield or, when the demand is considerable, the area burdened with the right must be formed into a separate circle to be worked solely with a view to furnishing the produce required.

A village has a right, every year, to 50 standing trees of a certain size and kind in one block only of a forest which is to be worked, we will suppose, by the selection method. The annual possibility of the forest, including the trees to be given to the right-holders, is calculated to be 500 trees a year. Therefore, in determining the size of the annual coupes, the coupe containing the area burdened with the right should be capable of furnishing the 450 trees once during the felling rotation, *plus* the 50 trees each year. That is, to say, if the felling rotation were 10 years, the coupe bound to furnish 50 trees a year ought to be capable of furnishing every 10 years 500 in addition to the 450 trees furnished by other coupes every 10th year.

Another way of providing for the right would be to set aside an area capable of furnishing 50 trees a year. This area would be formed into a separate circle. Such a course would, however, as a rule be impossible in mountainous country, where each small scattered forest has to furnish a few standing trees each year for the repairs of neighbouring hamlets.

In such cases the regulation may be best accomplished by sacrificing equality of yield, and by deducting the number of trees felled for the right-holders in a given area from the numbers to be felled in that area when its turn for exploitation comes round. Thus, suppose ten small scattered forests in the hills formed into fifteen annual coupes and worked by the selection method, the yield having been calculated at 500 trees a year. We will assume that the villagers in three neighbouring hamlets have rights in as many forests to 10, 20, 30 standing trees, respectively. Suppose that the prescribed fellings reach the forest No. 1 in the third year of the felling-rotation, 30 trees will have been felled ; so that, instead of felling 500 trees in this coupe, only 500—30 or 470 trees should be cut. Similarly, the felling reaches forest No. 2 after 10 years have elapsed. By that time 200 trees will have been felled in this coupe, so that only 500—200 or 300 trees should be removed ; and so on. A new calculation, based on a fresh enumeration, would of course be made at the commencement of the second felling-rotation, and the trees to be deducted would count from this date. Where the right-holders' timber was removed from areas subjected to selection fellings by cultural rules, the matter would be very simple ; as in such a case the state of the crop from a cultural point of view would determine the severity and nature of the felling, which would consequently be made light or heavy according to the quantity of material required by right-holders.

Extraction of dead or fallen trees.—As a rule, especially where it has been possible to choose a short felling rotation the dead and fallen timber may be left until removed with the ordinary exploitations in turn of each coupe. Where the local demand for produce is good, it may be advisable to allow the dead and fallen wood to be extracted annually or periodically. In that case, if the material to be removed by the fellings is prescribed, the dead or fallen trees extracted may, if desirable, be deducted from the permissible fellings as already explained with regard to trees felled by right-holders.

Works of improvement (other than cultural).—A working-plan would be incomplete if advantage were not taken of the study of the forest which it involves to ascertain and indicate the works of improvement, other than cultural, required. Very often the application of the plan necessitates the opening out of new roads for the extraction of produce, or the improvement of existing tracts and the construction of forest rest-houses; while the improvement of the boundaries, the better protection of the forest from fire or other injuries, the strengthening or re-distribution of the protective staff, may all require attention.

Such works and subjects should be indicated or discussed with whatever amount of detail may be required. They should not be vaguely suggested, as has been the case in so many Indian working-plans reports, but should be prescribed, though the exact time for their execution need not, and indeed often should not, be fixed. If new roads or new fire traces are required, their position should as a rule be shown on the map, and a rough estimate of their cost should be prepared in order to gauge the financial result of the whole working under the plan proposed. Very often it will be advisable to mark out on the ground the new roads or paths proposed, and to make use of these in laying out the coupes. The coupes, as we have seen, ought to be so disposed with regard to the roads that their produce can be readily and economically extracted. The only way to ensure that the coupes will be adhered to and the roads laid out in the manner required, is to mark out on the ground the lines of transport.

It may be useful to remark, with regard to works of this sort, that although included in the approved working-plan, separate sanction will be required for the necessary funds when the time for construction arrives. Another point, which will also be dealt with in treating of the control of plans, is that the works may often be only indicated as desirable, and need not be part of the permanent plan which cannot ordinarily be departed from.

Forecast of financial results of working.—The working-plan should contain a rough estimate of the revenue and the expenditure under the proposed working as compared with the actual results of the past. An attempt should in all cases be made to estimate the cost of proposed works of improvement.

When the areas to be felled have been decided upon, it will generally be necessary to make a forecast of the outturn of trees, or of material to be removed, so as to estimate the financial results of the proposed scheme of working. This estimate may be based either on an enumeration of the mature trees in the forest, on the previous outturn, or on experimental area fellings, etc. The enumeration should not, however, be mistaken for a *calculation* of the possibility, such as is made when fellings are prescribed by volume of material or number of trees.

CHAPTER IV.—THE WORKING-PLAN REPORT.

General remarks.—It is desirable, especially in India where frequent changes in the forest staff are unavoidable, that working plan reports should follow generally a fixed pattern. Uniformity not alone lessens the labour of mastering the contents of individual reports, but without it systematic scrutiny of the plans is well nigh impossible. The fact that each report deals with essentially different local conditions is no impediment to grouping and discussing together, under suitable headings and in logical sequence, the various subjects.

The general order to be followed in working-plan reports is indicated below. This is the arrangement prescribed at present by the Government; and it is only necessary to add that each report, while containing sufficient information to enable the soundness of its provisions to be tested and the ideas of its compiler to be followed, should be as brief and as simply written as possible. Lengthy descriptions and discussions of sylvicultural questions, mathematical calculations and complicated tables of statistics should, as far as possible, be avoided. The information deduced from statistical tables can generally be explained in a few words, and the application of mathematical formulæ in connection with Indian forests is liable to lead to erroneous conclusions. Some plans have never been applied, owing to the neglect of such rules. These monuments of misdirected energy represent, however, a large amount of labour, and, it must be added, a very considerable amount of expense, all of which, with the exception of the experience so brought, has been wasted.

Arrangement of the subjects.—A working-plan is a *forest regulation or prescription deduced from facts*; and, in order that the prescriptions may be intelligible it is necessary that the manner in which they have been arrived at should be explained and the facts from which they have been deduced stated. The working-plan report, therefore, naturally resolves itself into two parts; the first, containing a summary of the facts on which the proposals are based, and the second, a statement of these proposals—the working-plan proper—accompanied by whatever explanations are necessary in order to show why they have been framed

The nature of the subjects treated in each of these parts, and the order in which they should, as a rule, be arranged, are as follows :—

INTRODUCTION.

PART I.—SUMMARY OF FACTS ON WHICH THE PROPOSALS ARE BASED.

DESCRIPTION OF THE TRACT DEALT WITH.

Name and situation.
Configuration of the ground.
Underlying rock, and soil.
Climate.
Agricultural customs and wants of the neighbouring population.

THE COMPOSITION AND CONDITION OF THE FORESTS.

Distribution and area.
State of the boundaries.
Legal position.
Rights and concessions.
Composition and condition of the crops.
Injuries to which the crops are liable.

SYSTEM OF MANAGEMENT.

Past and present system, of management.
Special works of improvement undertaken.
Past revenue and expenditure.

UTILISATION OF THE PRODUCE.

Marketable products ; requirements to be met.
Lines of export.
Centres of consumption.
Mode and cost of extraction.
Net price of each class of saleable produce.

MISCELLANEOUS FACTS.

The forest staff.
Labour supply.

PART II.—FUTURE MANAGEMENT DISCUSSED AND PRESCRIBED.

BASIS OF PROPOSALS.

Working-circles ; how composed ; reasons for their formation.
Compartments ; justification of the subdivision adopted.
Analysis of the crop ; method of valuation employed.

METHOD OF TREATMENT.*

Object to be attained.
Method of treatment adopted.
The exploitable age.

THE FELLINGS.*

General scheme of working.
The possibility; how calculated.
Period for which the fellings are prescribed.
Areas to be felled annually or periodically; order of allotment.
Nature of and mode of executing the fellings; forecast of condition of crop at their conclusion.

SUPPLEMENTARY PROVISIONS.*

Cleanings, thinnings, or other improvement fellings.
Regulation of rights and concessions.
Sowings or plantings.
Roads, buildings and other works of improvement common to the whole area.

MISCELLANEOUS.

Miscellaneous prescriptions.
The forest staff; changes (if any) proposed.
Forecast of financial results of proposed working.

APPENDICES.

Maps.
Description of the crop in each sub-division, written or graphic.
Record of valuation surveys.
Record of observations of rates of growth.
Miscellaneous statements.

The introduction.—This should briefly explain the time occupied in the preparation of the plan, the establishment employed, and the expenditure incurred under each head; any special difficulties encountered, as well as other facts which merit permanent record but which do not find a place in the body of the report.

Example.—The field work, which included the survey of the forest to be exploited under this plan, was commenced in the beginning of June 1884, and continued, with a short intermission during the rains, until December, when travelling in the hills became impracticable. It was recommenced in the following May, and brought to a close at the end of the same year, 1885.

The establishment employed during the first season consisted, in addition to the Officer in charge and his clerical staff, of two native sub-surveyors, lent from the Forest Survey Branch towards the end of the season, and of a small staff of khalasies, recorders, tree-measurers and coolies. During the second season two Forest Rangers were employed in addition to the temporary staff entertained in the previous year.

The total cost of the work, including the pay and allowances of the Officer in charge, the subordinate staff and all other charges, amounted to R9,400, or to R70 per square mile. The high cost-rate is partly due to the rugged nature of the country in

---

* Each working-circle should be separately dealt with as regards the method of treatment, the fellings, and all supplementary provisions except those common to the whole area, such as road construction.

which the forests are situated and the consequent difficulty of travelling; but it is also, in a great measure, due to the employment of temporary subordinates, and the lateness of the season when work was commenced in the first year.

Name and situation—The name of the tract dealt with in the report, generally some forest charge, and the civil district or territory in which it is situated, together with that of the Forest Division to which it belongs, should be stated; and it should be explained whether the wooded area comprises one or several separate forests. The vicinity of large towns or markets for the produce, or of rivers, roads, or railways leading to these places, should be very briefly mentioned here.

Example.—These plans have been prepared for what are known as the "Naini Tal Forests" which comprise twenty separate areas situated in the Kumaun District, in the *pattis* of Chakrata Pahar and Dhangakot, within the jurisdiction of the Commissioner of Kumaun. The forests surround and supply with produce the Municipality and Cantonment of Naini Tal, and constitute one of the two ranges of the Naini Tal Forest Division.

Configuration of the ground.—It should be explained whether the forests are situated in hilly or level ground, or on a plateau; whether they form part of one or several river basins; at what height or heights above the sea they are found, and what their relative position is with regard to the surrounding country.

Example.—The forests are situated on comparatively low hills, forming a series of superimposed terraces or plateaux with sloping sides. The tops of the hills are flat; so that, instead of ascending to a ridge, we ascend to a plateau, and the action of time has not been able to destroy the original terrace-formation which gives to the hills their characteristic appearance. Precipices are, however, uncommon, and the slopes are usually easy enough for a saddled horse to be led almost everywhere. The hills rise from 1,000 to seldom more than 1,700 feet above the deepest valleys. The forests, as might be expected, occupy the upper plateaux and slopes of the hills, cultivation having taken possession of the lower slopes. The main ridge occupies a continuous sinuous line, extending from south-west to north-east, forming the water-parting between the Ramgunga stream on the north and the Pahdi river on the south; but the greater portion of the area lies within the drainage basin of the latter river.

Underlying rock and soil.—The general character of the geology of the country, the resulting soils and their relation to the composition of the forests, should be explained. The explanation should, however, be brief and of a general nature; as a more detailed description of the soils in each block is, when necessary, separately given.

Example.—The entire rock-system belongs to what is known as the *Deccan traps* and is consequently of volcanic origin. The soil resulting from the disintegration of these trap rocks is a red ferruginous loam, fairly fertile where deep. This, however, is not the case on the level surfaces of the plateaux where disintegration is slow and where, in some cases, the unchanged rock protrudes at the surface. In other parts the underlying rock, being everywhere laminated, compensates to some extent for shallowness; but generally speaking, the soil is not favourable to tree growth, as it is too superficial and dry. On the slopes, however, which separate the various plateaux and along the bottoms of the numerous gullies, the soil, formed of deep accumulations and resting on a sub-soil of loosely disintegrated trap, is well-drained and admirably suited to tree-growth. An intrusion of granitic or gneissic rock, about three miles broad crosses the area from south-east to north-west, and the resulting soil is a fine micaceous sand, which of itself is usually unfertile for agriculture, requiring heavy manuring, but which, under broad-leaved forest, becomes fairly and even highly fertile.

Climate.—All that is generally required is a simple statement of the facts with most of which every cultivator in the locality is practically acquainted. Periods of drought or of excessive rain, of great heat or of cold, frosts, dangerous winds and the like which have a notable effect on the forest vegetation or on fire-protection, sowings, plantings or other sylvicultural operations, should be briefly explained. A few remarks may be usefully added regarding the healthiness or the reverse of the climate, when this question affects the proper execution of work or the duties of the establishment generally.

The climate and soil together represent the productive capacity of the area or locality and determine the species and method of treatment best adapted to the results required. The character of the climate depends on the situation of a given locality, and is described by stating the different local peculiarities of the atmosphere as regards temperature, degree of moisture, prevailing winds, etc.

The points to be borne in mind when it is desirable to report in detail upon the climate of a given locality are, therefore, as follows :—

(1) Geographical situation.
(2) Height above sea-level.
(3) Relative height and position with regard to the surrounding country.
(4) Slope, aspect, and topographical features generally.
(5) Temperature, mean, average, and at different seasons; periods of greatest heat and cold and their duration.
(6) Species which thrive in the locality and their peculiarities as regards climate.
(7) Usual state of the atmosphere, whether clear or the reverse, and its dryness or humidity at different seasons.
(8) Rainfall, mean, annual, and at different seasons; drought and its duration; snow-fall and its duration.
(9) Winds and storms, their duration and force, damage done by them.

Points (1) to (4) are of relatively small importance in the plains; but circumstances of climate should be applied to the questions at issue and be explained in a popular manner.

Example.—The distinguishing characteristics of the climate are its dryness and the great divergence between the temperature at different seasons of the year. These extremes of heat and cold, combined with the dryness of the atmosphere during most of the year, render the propagation of all but a few species, such as *sissu*, impossible ; and even the hardy *kikar* is killed off by frost unless artificially shaded in winter. The monsoon rains nominally commence in July and continue to the end of August; but they are neither very constant nor regular, and sometimes almost altogether fail. Plantings and sowings are only possible during this short period, and even then often fail in consequence of the dry hot weather during September. As a rule, a large proportion of the annual rain falls during the winter, showers commencing at the end of December and lasting to the second week of January. The months of March, April and May are more or less stormy as well as hot and dry. The dangerous season for fires is thus much prolonged; but, on the other hand, the grass withers early and can be burnt with safety. Easterly winds prevail but are seldom violent, and their effects can be neutralized to a great extent by protective bands of trees along the west boundary. The climate is, on the whole, healthy, even during the scorching dry heats, except in the deeper valleys and irrigated tracts. Owing to the dampness of these latter places the forests remain green throughout the year, and are so unhealthy that labourers cannot be induced to reside or work in them during the warm season from May to October.

gricultural customs affecting the proposals and wants of the neighbouring population—In most instances the agicultural customs and the mode of life of the local population have an important bearing on the management of the forests, both with regard to the direct supply of the wants of the people in forest produce and to the adoption of a system of management that will interfere as little as possible with established customs. Such facts, where they influence forest management should, therefore, be briefly stated.

Example.—The inhabitants are all of necessity cattle-farmers as well as cultivators, as there are no canals in the district and water lies too deep from the surface to make well-irrigation profitable. Hence cultivation is confined to the areas near the rivers, where, however, owing to the irregularity of the floods, farming affords at best but a poor and precarious means of subsistence. The combined system of cattle-farming and tillage in vogue is, therefore, the only means by which the people can manage to support themselves. Even in the driest years, when the bare lands near the villages contain no fodder at all, there are in the forests some grasses and the leaves of many trees which can be used as fodder, and by these means the villagers contrive to keep large herds of camels and cattle, and flocks of sheep and goats fiom which they derive a considerable addition to their income and their food-supplies. It follows that, while grazing must be provided for, the number of animals pastured should be restricted to the possibility of the forest in leaf and grass fodder. Unless such a limitation is enforced the forests must inevitably disappear; and the great increase in the stock kept by the villages threatens this already.

Distribution and area.—It is generally necessary, especially where the report deals with a number of scattered forests, to explain in a few words how the gross area is distributed.

Example.—The total forest area is formed of a large number of separate blocks, occupying the steeper slopes and less accessible portions of the higher hills, all cultarable portions and the lower valleys having been brought under the plough. The number of demarcated forests in the valley is 80, and the average area of each is about 900 acres. These relatively small forests are scattered over the whole area of the valley, in the basins of the tributaries of the Ravi and Siul, at elevations of between 5,500 and 8,500 feet above sea-level.

The detail with which the area statement should be compiled depends on the nature and degree of elaboration of the plans to be prepared. When possible, the areas of the wooded, blank and unculturable portions of the forests, and the areas occupied by enclosures of private or other land not under forest management but within the boundaries, should be stated *separately* for each block or sub-division. This information is required in laying out coupes, especially where the possibility is prescribed by area. Where, however, large areas are to be felled over by the selection method a less complete area statement suffices, but would usually be supplemented by a more detailed description of the blocks and of the wooded, blank and unculturable area in each.

In addition to stating the area, it should also be explained how and with what degree of accuracy the forests have been surveyed, what maps exist, and how the acreages have been ascertained.

Example.—The area of the working-circle can only be given approximately, as accurate maps are available only for the Deota forests.

Forests.	Blocks.	Area of each block in acres.	Total area (acres).	Remarks.
Deota .	Lambatah . Deota . Bamsu . Saras . Kotigad .	2,883 2,937 2,323 1,234 1,105	10,482	The Deota and Chansil forests were surveyed in 1881-82 by the Forest Survey Branch, and the areas given are those furnished by that Branch as calculated from the maps on the scale of 4″=1 mile.
Chansil	Chansil . Kotigad .	1,805 3,968	5,773	For the remaining forests which were demarcated in 1885, only rough sketch maps on the scale of 2″ = 1 mile exist, and the areas given are subject to correction when the regular survey, now in progress, has been completed.
Sahlra .	...	...	1,410	
Noranu .	...	...	1,520	
Naintwar	...	...	3,450	
Datmir .	...	...	2,560	
Total	...	...	25,195	

The following is an example of a more detailed statement. Very often, however, it will be necessary to indicate not merely the total *wooded* area but also the area of each type of forest in each block. Where a stock map has been prepared this offers no difficulty :—

Blocks.	Area.								Remarks.
	Wooded.	Blank.	Forest roads.	Forest buildings.	Other forest lands.	Total under Forest Department.	Area not under Forest Department.	Total area within boundaries.	
Dhamia	2,643.4	123.8	111.1	11.6	...	2,894.9	116.4*	3,011.3	* 13.2 acres occupied by public roads ; 1.2 acres in corner of compartment No. 1 by Canal buildings ; 81.0 acres by lands adjoining the Canal, and 21 acres by the Railway.
Paimar	1,699.3	68.0	83.0	53	16.0	1,871.6	59.0†	1,930.6	† 38.0 acres occupied by the Canal lands, and 21.0 by the Railway.
Kote	1,746.2	274.8	21.7	1.6	3.6	2,047.9	61.5‡	2,109.4	‡ 38.5 acres occupied by Canal bungalow and compound and 23.0 acres by lands adjoining Canal.
Lurla	2,310.2	91.1	138.0	...	...	2,539.3	101.0§	2,640.3	§ 101.0 acres of lands adjoining Canal.
Total	8,399.1	562.7	353.8	18.5	19.6	9,853.7	337.9	9,691.6	

Boundaries.—In order to justify proposed works of improvement, or the absence of any proposal of the kind, with regard to the demarcation, it is necessary to ascertain and state whether the boundary marks are sufficient and suitable or not; whether they are in good or bad condition; and whether they are well placed. In some cases a re-demarcation of parts of the boundary may be required, and, if so, this should be justified. The nature of the surrounding properties, the likelihood of trespass, and such other points as bear on the size, degree of proximity, etc., of the boundary marks should be dealt with in such detail as may be necessary in each case. A "register" of boundary marks should not usually find a place in a working-plans report.

In many cases, for instance where the annual clearing of the boundary lines for fire-protection or other purposes is necessary, it may be advisable to state the length of lines to be kept clear or the number of marks to be repaired, etc.,

The boundaries of private lands, included in the forest, should be described in the same way as the external boundaries.

Example.—The boundary marks used along the outer boundaries are substantial and sufficient. The whole of this boundary is demarcated by square masonry pillars 3′ ×3′ × 4′; whilst pillars, also square, measuring 2′× 2′ × 2½′, mark off the boundary between the open forests and those free of rights. There are in addition 318 round pillars, 4½′ in girth by 2½′ high, demarcating the chaks or village lands within the forests. The shape of the boundaries is, however, very defective, a crooked or curved line between two consecutive pillars being common. The consequence is that it is extremely difficult to follow the boundary, and encroachments might long pass unnoticed. For the reserve this drawback has been overcome by erecting a ring fence of rough posts; but something should be done as soon as possible to render the boundaries of the other forests easily and permanently recognisable. Otherwise harassing disputes in the near future are inevitable. The adjoining estates consist of village lands, the cultivation of which not infrequently is conterminous with the forest boundary.

It may occasionally be advisable to prefix to the description of the boundaries remarks explanatory of the circumstances under which the demarcation was made.

Enclosures within a reserve are often of such importance, as regards the management of the forest, that it may be necessary to notice them in detail.

Legal position of the forests.—A brief account of the manner in which the forests were acquired, and how they are held by Government, of their settlement (if one has been made) and of their present legal *status*, should be given. The Act and section of the Act under which the forests have been declared "reserved" or "protected," as well as brief particulars of all important orders of the Local Government concerned bearing on the *status* and management of the areas, should also be indicated. Special regulations affecting the forests, such as grazing rules, should be mentioned, and their bearing on the legal position of the tract should be explained. This section of the report should not, however, be burdened with details regarding the past management of the forests. This subject will be separately dealt with in another section.

Example.—Jaunsar originally formed part of the Sirmur or Nahan State. In 1815 it was conquered by the British, since which time it has been under British rule. But it was not until 1860 that the right of Government to the ownership of all waste land was enforced by rough limitation of the rights of each collection of village communities. It was clearly laid down in this settlement that the rights of the villagers in the forests consisted in grazing, in collecting dead wood for fuel, and in cutting timber for their own use only.

In 1865-66 the proposal to establish a military cantonment at Chakrata gave rise to the issue of different orders for the protection of the more valuable tracts of forest, and in 1868 a Forest Officer was appointed to take charge of them. In 1869 the demarcation of certain portions as Government forest was directed, and three classes of forest were formed:—

First class.—Forest areas practically free of rights.

Second class.—Forest areas under the control of Government, but subject to certain rights of usage.

Third class.—Forest or waste land, the use of which was allotted to the different *Khats*, the right to sell produce only being prohibited.

In 1878, on the Indian Forest Act (VII of 1878) coming into force, a settlement, under Chapter II of that Act, was made of the first and second class forests, and such areas as could not be secured free of all rights were included in the category now called *Unclassed State Forests*. The settled forests, which it will be seen are free of all rights, were gazetted as " Reserves " under Section 19 of the Forest Act (*vide* G. O. No. 408 A. of 1st August 1880). The unclassed forests are not (*vide* G. O. No. 77 F. C. of the 13th June 1871) managed by the officers of the Forest Department, the enforcement of the limitation as to the sale of produce from them being left to the civil officers.

Rights and concessions.—It is generally necessary to explain the nature and extent of the rights with which the forests are burdened, so far as they affect management and sylvicultural treatment. But it is only where definite proposals are made for commuting particular rights, or for providing for them in some special manner, that full details of each right need be given. And if a detailed report is required it should be prepared separately from the working-plan.

It should be clearly understood that the body of a working-plans report is not the place for a detailed *record-of-rights*, the up-keep of which is separately provided for in the Forest Code. As, however, it sometimes happens that the rights, although legally settled, have not been recorded in a convenient form, it may be advisable in the preparation of the working-plan to redraft, under proper authority, this record in a more intelligible form for ready reference. If so desired, the record may often be suitably appended to the working-plans report.

Example.—The forests in this Division come under the action of the third clause of Section 34 of the Act, and, no settlement or enquiry having yet been made, the existence or non-existence of rights unfortunately still remains an open question. The following concessions have, however, been made by administrative order in favour of the inhabitants of certain villages in the immediate vicinity of the forests, *viz.*:—

Sattikbala	. . . .	10 chil trees 2 feet diameter, annually.		
Malan	. . . .	20	,, ,, ,,
Laldhang	. . . .	35	,, ,, ,,
Amgadi	. . . .	26	,, ,, ,,
Bedasni	. . . .	63	,, ,, ,,
TOTAL		154	,, ,, ,,

The exercise of these rights is restricted to the actual domestic and agricultural requirements of the villagers in whose favour they are recorded, and they cannot be exercised for purposes of trade. The maximum size of a house, for building and maintaining which timber can be claimed, is 40 feet × 36 feet × 16 feet.

Any trees, except the protected kinds, may be lopped for green leaves; but the lopping is limited to branches not exceeding 2″ in diameter at the trunk.

The grazing allowed under the orders above mentioned is as follows :—

Tract affected by the concession.	Total head of cattle admitted to graze at half rates.					Remarks.
Name of block.	Acreage.	Buffaloes.	Cows and bullocks.	Horses, ponies, and donkeys.	Sheep and goats.	
Malan .	2,136	38	4×5	...		The present full annual rates per head of each class of cattle are, respectively, 4, 2, 1, and ½ annas. Hence the total money-value of the concessions at the present rates is R 2,9.5.3 per annum.
Amgadi .	3,135	223	1,165	52	175	
Bedasni .	6,304	106	1,434	33	220	
Total .	11,575	367	3,114	85	395	

It is always well to estimate the quantity of timber or fuel for which provision must be made. This information is necessary in order to arrange for the exercise of the rights and for their regulation which, as will be seen, may necessitate the grouping into separate working-circles of the areas affected. The bearing of the rights on the management of the forest should be noticed, and it should be pointed out in what way the rights should be regulated. An endeavour should always be made to estimate the value of the concessions granted.

Composition and condition of the crops.—As a detailed description of the crops in each block is usually separately given or figured in the stock map, this section should contain only a broad general description, special attention being paid to points, such as the following, not dealt with in the detailed description :—

The different classes or types of forest and their distribution ; principal and accessory species, and their relative proportions.

The condition of the crop ; the dominant age-classes ; the quality and density of the growth ; absence or presence of blanks or glades.

Reproduction ; the presence or absence of natural seedlings, with explanatory remarks.

Example.—Taking the forest as a whole, the stock may be generally sub-divided into three distinct zones which roughly occupy the following areas :—

(1) *The Karshu oak zone.*—Beginning at about 9,000 feet and extending to the highest point of the basin. It comprises pure or mixed open or dense patches of the silver and spruce firs and of *karshu* (oak), interspersed with bird cherries, maples, service trees and yews. In this zone, wherever the ground is fairly level, we obtain extremely rich pastures Deodar just enters the zone.

(2) *The deodar zone.*—From 6,000 feet to 9,000 feet. Deodar is spread throughout this zone. The tree does not, however, grow pure, being mixed with a larger or smaller proportion of spruce and *moru* principally, and silver fir exceptionally. It is most abundant, and generally constitutes the predominant tree, between 7,000 and 8,500 feet of elevation. Above and below this belt deodar confines itself almost exclusively to ridges and well-drained slopes, particularly the former.

(3) *The chir zone.*—From 6,000 feet to the lowest altitude of the circle (4,416 feet). The predominant type of forest is broad-leaved, consisting of *ban* (oak) and a few horse chestnuts, maples, etc., at the higher elevations, and chir with small trees and shrubs elsewhere. Numerous grazing grounds exist or recently existed in the *ban* area, which is therefore interspersed with comparatively extensive plots of open low scrub.

The area producing deodar, which is the one marketable tree, is about half the total area of the forest. The crops are all irregular, no distinct age-classes of gradation of ages existing; and every variation in density is to be met with, from open grassy blanks to dense mixed deodar and fir forest. In the following statement, the areas occupied by each of these zones of growth are recorded :—

Zone.	Wooded.	Blank.	Unproductive.	Total.
Karshu	2,795	1,024	2,108	5,927
Deodar	7,855	1,603	1,002	10,460
Chir	1,034	1,038	428	2,500

In the deodar zone, save where the cover is too dense, reproduction is good, seedlings of deodar as well as of kail and rai being plentiful. In the Palangi block especially the young growth of *moru* (oak) is excellent. In those blocks which have been worked over natural reproduction is exceedingly satisfactory.

With what degree of detail the principal species should be described depends on the special circumstances of each case, and must therefore be left to the judgment of the working-plans officer. But, broadly speaking, with regard to every important species of which the treatment is not well known, it will be useful to record whatever information can be collected on the following subjects :—

> Characters of the soil and of the locality in which the tree is found growing and in which it thrives best.
> Requirements or peculiarities with regard to reproduction, seedling, etc.; aptitude to produce shoots or suckers.
> Growth in girth and height; size attained; longevity.
> Most suitable method of treatment; peculiarities as regards supporting shade, or the reverse.
> Products yielded and purposes for which employed; qualities of the wood.

Information concerning the origin of the crop or important facts connected with its past treatment often throws light on its present condition. In the case of plantations the origin and past history can be readily described, and for many other crops also much may be learned.

Example.—The large size of the few old trees still remaining would of itself be sufficient to prove that a fine forest growth at one time covered at least the more favourably situated areas, and there is besides clear evidence that nearly all the old growth has disappeared through *jhuming*. The fact is that the level tracts, which typify these hills, attracted settlers at an early date after the pacification of the country under British rule; and, although *jhuming* did not extend far up the hill sides which are generally too steep for this mode of cultivation, the trees did not escape injury. The best kinds were exported to the plains, being generally felled four or five feet from the ground, and young shoots were similarly hacked off at soon as they attained a saleable size. The inferior species were lopped to provide fuel or fodder, while the entire area was overrun by herds of cattle and was subject to annual fires. The evidences of this ill-treatment are very visible on the crop which now remains.

Injuries to which the crops are liable.—It is only necessary, as a rule, to explain those causes of injury, such as breaches of the forest regulations, fires, grazing, climbers and other injurious growths, depredations by noxious animals, etc., which in practice may be prevented or restrained, and which have an important direct bearing on the forest management. Purely scientific information with regard to fungoid growths or insect pests, however interesting or even useful it may be, is generally out of place in a working-plan. The injuries may be conveniently classed under two main heads, *viz.*, those due to natural causes, and those caused by man.

Example.—The chief preventible causes of injury are grazing and fires. Many other injurious customs formerly prevailed, such as tapping for resin, taking large strips of bark off the trunks of fir trees for roofing temporary huts, and hacking green deodar and other conifers for fuel or torches; but these malpractices have been altogether put a stop to.

The bad condition of the forest as regards reproduction is largely due to the very irregular manner in which grazing is allowed. The cattle graze all the year round at a few favoured spots near the villages. There are thus numerous congested grazing centres scattered throughout the forests, whilst there are also considerable areas where grazing is very slight or scarcely takes place. The closing and opening in regular rotation of different blocks, and the consequent distribution of the grazing over the entire area would, in a great measure, mitigate this evil.

The chief cause of injury is, however, fire. With the exception of the protected blocks, only some damp, shady hollows, covering an insignificant area, escape, and the general effect on vegetation is deplorable. The quality of the trees is inferior, a large percentage of the stems is hollow, and saplings are often burnt down to the ground. The greater portion of the stock is really derived from coppice shoots; while in other respects reproduction is almost entirely checked owing to the seeds and seedlings being destroyed. The density of the crops is also sensibly affected, the formation of large blanks is increasing, and the spread of bamboos and grasses, such as babar, has been encouraged rather than retarded by these fires. Considering that there is a certain demand for wood of all kinds in the range, it is of course important to extend fire-protection over a larger area, and this may gradually be done without causing hardship or even local discontent.

Climbers in the upper forests, unlike the *malján* and others in the sál areas do little or no damage. The commonest of the climbers and creepers are ivy, vines, particularly *Vitis himalayana Schizandra grandiflora* and *Ficus-scandens.* These perhaps do more good than harm by keeping the trunks moist and by killing of the lower branches. No measures are, therefore, necessary with regard to them.

Past and present systems of management.—The system or systems of management which have been followed in the past should be briefly discussed, with a view to preventing a repetition of such mistakes as may have been made, and so that the present condition of the crop and the system of management actually in force may be better understood. This may necessitate some historical remarks brief enough to render the explanation intelligible.

Example.—During the first three years of British rule all forest dues were leased out to contractors with the transit dues on merchandise, and subsequently they were farmed out to the zamindars of the parganas in which they were collected. In 1844, in consequence of the difficulties regarding boundaries that had occurred, the collection of these dues was entrusted to the authorities of the district; but two years later the duty of collecting them was restored to the zamindars. In 1866 the reservation of the lands was recommended, and a proclamation was issued prohibiting the cutting of the sâl within these areas; and thus the system of Government protection commenced. Since that time the following systems of management have been attempted, with, as will be seen, varying results.—

Leasing out for one year definite areas for the extraction of produce.—This system saved the establishment much trouble, as the lessees issued their own passes and exercised their own check; and since they understood the requirements of the local markets better than the forest officers, they probably paid Government a larger sum than would have been collected in detail by the Forest Department direct from petty purchasers. But the method led to collusion between the lessees and the subordinates intended to supervise them: trees were killed standing, and timber of forbidden descriptions was removed. The system has now been abandoned.

Removal of produce under passes issued at revenue stations.—Generalized too far, this system—which was enforced during some years after the abolition of the first—leads to the very same defects; but, restricted to the removal, under passes valid for a short period, of small fallen wood and of marked and girdled trees fit only for fuel, is a very desirable adjunct to other methods of workings.

Sale of standing trees by the Forest Officer by public auction or by private agreement.—This has been found by far the best system in disposing of large quantities of valuable produce and should be extended as far as circumstances will permit.

It will generally be necessary to explain fully and to criticise the system of management in force, pointing out defects observed in it in justification of any changes that may be proposed.

Example.—Under the system of working in force—

Green standing trees may only be cut after marking and with the previous sanction of the Divisional officer. An exception to this rule is sometimes made in the case of inferior species.

Fallen green trees may only be removed under the pass and mark of Foresters in charge of ranges or beats.

Bamboos may only be cut in specified areas and under permits issued by Range Officers.

Purchasers are allowed to enter the forests and to remove the produce they require on taking out a pass at any of the revenue stations and on paying the dues according to a fixed tariff. The general rule is, however, subject to the following restrictions:—

That permission to work be obtained before the forest is entered, and that the forest rules be observed.

That cattle shall not be brought into the forests without the grazing dues having first been paid, or remain in the forests save at authorized cattle-stations.

Although there can be no doubt that more timber and firewood are produced over the entire area than the quantity sold annually, it appears that, on the other hand, some of the more accessible parts have been overworked, whilst large but less accessible areas have remained untouched. The rules of working require to be revised so as to remedy this.

Special works of improvement.—In most cases there will be works of improvement or construction for the amelioration of the crop or for the extraction of produce, such as sowings or plantings, timber slides, roads, or bridges; and occasionally special operations may be required, such as drainage or irrigation works, barriers to prevent the erosion of the banks of streams, and so forth. All these may be described in this section in such detail as circumstances require.

In discussing cultural or other works of improvement undertaken in past years attention should be drawn to important results of experience so acquired. But if it is necessary for the information of the local officers to describe these works in detail, this should be done in a special report or in an appendix to the working-plan.

Example.—It has been ascertained from the records that plantations of deodar, walnut and ash have usually succeeded; while direct sowings have uniformly failed. Even in the nurseries the germinating seedlings have survived with difficulty, but once they are a few months old they have given no trouble. The cost of planting out from the nurseries has, on the average, amounted to R10 per acre, or, including nursery charges, to R18 per acre of established transplants.

Past revenue and expenditure.—Where the figures are available, the average receipts and expenditure for the past few years in connection with the whole area dealt with and, if possible, regarding each class of forest in this area, should be given. If no separate accounts are available an estimate may be made. The prices realised for the produce, exclusive of cost of extraction and of establishment employed, should, as far as practicable, be stated. These prices should be discussed, use being made of the figures to justify proposals, if any are made, for increasing or reducing expenditure under these headings.

Example.—From a statement showing the revenue and expenditure during the past ten years, compiled from the records in the Range offices, it appears that the average annual gross receipts during the past ten years were R1,64.000, while the expenditure on timber works was R44,000. The profits derived from the forests were therefore R1,20,000 annually, and of this sum there has been expended—

On roads and buildings	2 per cent.
On cultural improvements	3 ,,
On establishments	2 ,,
Total expenditure .	26 ,,

It is beyond dispute that this expenditure is far too low. The condition of the forests requires the expenditure of large sums annually on cultural improvements; while the low net price realised (only two annas per cubic foot on an average) for the produce shows that new lines of extraction are indispensable in order to place the produce at a reasonable cost where it is wanted. An expenditure of probably 25 per cent. of the profits should be devoted to improvements for some years to come.

Marketable products.—It should be stated what products of the forest are saleable, the purposes for which they are used, and the quantities of each sort consumed by different classes of the community or required to meet the general demands of the public. In preparing the plan a statement should, therefore, be tabulated from the records showing the quantities of produce exploited by Government agency, sold to purchasers direct, used departmentally, granted or sold at privileged rates to right-holders and given as free grants.

Example.—The following statement exhibits the average quantities of produce removed annually during the past five years:—

PRODUCE.	AVERAGE QUANTITIES OF PRODUCE REMOVED OR UTILIZED ANNUALLY.								REMARKS.
	By Government Agency.		By Purchasers.		By Free-Grantees.		By Right-Holders.		
	Quantity.	Value.	Quantity.	Value.	Quantity.	Value.	Quantity.	Value.	
	Cub. ft.	R	Cub. ft.	R	Cub. ft.	R	Cub. ft.	R	The material removed has been recorded in various ways, by trees, mds., etc. The measurements have been reduced to cubic feet by estimate.
Green sal timber	141,500	16,800	4,216,000	48,000	6,000	720	15,000	1,500	
" salu "	69,000	9,020	200,000	24,000	...	...	...	...	
" other woods	...	...	108,000	4,280	...	...	...	...	
Fuel	20,000	625	180,000	23,000	...	...	20,000	2,400	
Bamboos	...	...	...	5,000	...	...	...	...	
Grazing	...	...	...	2,937	...	...	...	790	
Miscellaneous	...	...	...	384	...	...	...	...	
Average 5 years 1885-86 to 1889-90	...	26,445	...	107,801	...	720	...	4,690	

Lines of exports.—The roads or main export lines, passing through the forest or in its vicinity, by which the produce is conveyed to the markets where it is consumed should be mentioned, and the state of repair and adequacy of these roads should be discussed. Rivers and streams used for the extraction of produce should be similarly described, and any necessary improvements of the waterways should be suggested.

Example.—The lines of export leading from the forests are the Sieul River, on which are situated the villages of Oderpur and Dharampur, and the main cart-road leading to the city of Rampur. In the Sieul itself, at Bagti, a narrow channel is so blocked up by a series of large boulders that it is difficult to float logs from one end of the passage to the other. Blasting operations are required here in order to free the channel.

The carting road is in the charge of the District Engineer, and its repairs are paid for from District funds.

The following roads, which are the main arteries of the Range, have to be kept in good order so as to be passable for mules :—

Misan to Sagti . . . . 12 miles.
Dand to Sooli . . . . 6 ,,

New roads of the same sort are required in order to open out the Sain forest: length from the forest to the cart road, 8 miles.

Ordinary roads or paths, when no longer wanted for the transport of timber and fuel, need not generally be kept in repair, but will not altogether disappear and can be restored when again required.

Markets.—The size and importance of each of the centres of consumption supplied from the forests under examination, their distance from the forest and the produce consumed in each, should be discussed with such detail as appears necessary.

The following heads under which the facts may be recorded should be borne in mind :—

Name of market.
Distance from the forest.
Line of export.
Description and quantity of produce consumed.
Sources from which supplied.
Quantities coming from Government forests.
Rates paid by dealers.

Mode and cost of extraction.—The manner in which the produce is extracted should be explained, and the cost of felling, transport, etc., should be given. Improvements will of course often be suggested with the object of reducing the cost. The establishment of new roads and the improvement of existing lines of export may be justified in this section.

Example.—All the produce reaches the market by railway, to which it is conveyed from the forests distant from 2 to 6 miles by bullock cart. The rates charged by the Railway Company are given below. The cost of conveyance to the line by bullock cart amounts on an average to six pies per maund per mile. This is the chief expense in the extraction of the produce, and in a separate report it has been proposed to substitute carriage by tramway for the present system. A tramway will reduce the cost of extraction three pies per maund per mile. The cost of constructing the proposed tramway, in accordance with the estimates given in the detailed report, has been included in the financial forecast.

Net price realised for the produce.—Upon the proper consideration of this subject often depend the exploitable age and the method of treatment to be applied. The report should state separately for each class of produce, the purposes for which it is used, the net revenue realised after deducting all costs of felling and extraction, etc. An example of the calculations which may be necessary has been given in Chapter II in discussing the exploitable age. The result arrived at may be stated in words or in tabular form as most convenient.

Example.—The following statement indicates the gross and net prices realised for each class of produce:—

Description of produce.	Gross price per cubic foot (solid).	Net price per cubic foot (solid.)	REMARKS.
	R	R	
Trees standing in forest	0·50	0·50	These figures are calculated from the average sales and the expenditure incurred during five years.
Fuel billets, 2' 6" long, 10" to 2" diameter.	0·66	0·23	
Fuel billets, 5' long, 2" to 1" diameter.	0·82	0·06	

The forest staff.—The strength, duties and cost of the forest staff should be stated, and the adequacy or insufficiency, as the case may be, of the existing establishment commented upon. Where alterations are necessary this should be stated and explained.

Example.—The following ranges and beats have been established :—

Range and head-quarters.	No. of beat.	Forests or blocks included in each beat.	Area of beat.	Head-quarters of beat.	REMARKS.
			Acres.		
Deota	1	Partil, Temple, Deota, Katatach.	1,798	Deota.	The charge of the range is at present held by a Forester.
	2	Tadiar and Bagiar	1,139	Tadiar.	
	3	Pipal, Pamsu	1,037	Pamsu.	
	4	Dhikuri	1,930	Suhlra.	
	5	Naintwar and Datmir	1,010	Datmir.	
		TOTAL AREA	6,914		

The duties of the subordinate establishment are particularly heavy, as the timber works are conducted departmentally. The injuries to which the forests are liable have already been described, and the protective duties of the guards will be understood. The average area of each guard's beat is about 1,400 acres; but the forests are scattered over a large area of exceptionally rugged country. It is often difficult to obtain suitable forest guards, as the hill men are unaccustomed to discipline of any sort. Proposals for increasing the staff, and for certain changes in its disposition, will therefore be made.

Labour supply.—It should be said to what extent and at what rates of payment it is possible to procure local labour, and whether, at particular seasons of the year, there are difficulties with regard to the supply. Any other remarks that may appear desirable in connection with the execution of works in the forests should also be recorded under this head.

PART II.—FUTURE MANAGEMENT DISCUSSED AND PRESCRIBED.

Working circles.—In explaining what working-circles are proposed, their formation should be fully justified with reference to :—

The state of the crop and method of treatment to be applied.
The position of the natural land-marks.
The demand for the produce, and the most desirable size for the coupes.
The administrative charges.

It will generally be convenient, specially when there are a number of working-circles, to exhibit in a tabular form the areas comprised in each circle together with the names of the forests or blocks concerned.

Example.—It will have been seen that the forest area consists of two portions, separated from each other by the main road from the railway line to the tahsil, and perfectly distinct in the character of the crops and in legal constitution. That portion situated to the east of this road has an area of 13,631 acres, is reserved and contains, as we have seen, an irregularly coppiced forest of sal; while the area lying to the west of the road is unclassed forest, and contains only scrub jungle which it will be proposed to work for grazing. This latter block must, it is evident, form a separate working-circle, and it is proposed to call it the "grazing circle."

As to the first-mentioned block, it is too large to form conveniently a single circle, and it is, therefore, proposed to divide it into two, separated by the stream and the path leading to the village. But both circles will be simultaneously worked to supply the same market which is fed by the railway. It may be argued that this division of the eastern area into two circles will render the working-plan more complicated. The cost of extracting the produce will, however, be decreased by reason of the lesser distance it will have to be carried; while the products of the fellings will not be so difficult to extract and dispose of. The following area statement shows the distribution of these working-circles, called the "eastern" and "western," respectively. Each will conveniently form a Forester's charge, replacing the present arrangement by which a Ranger has charge of the reserved forests only, and two guards, belonging to the sub-division but not under the Ranger, have charge of the grazing area:—

	Working-circles.	Blocks included.	Area in acres.	REMARKS.
Eastern		Mangir	328	
		Bhandal	1,180	
		Langers	1,302	
		Malla	1,448	
		Himgri	1,339	
	TOTAL AREA		5,565	
Western		Barnota	1,250	All these forests are reserved.
		Sai	1,246	
		Alwas	728	
		Batra	1,420	
		Tisa	746	
		Chanju	1,483	
		Kalel	1,188	
	TOTAL AREA		9,066	

Working-Circles.	Blocks included.	Area in acres.	Remarks.
Grazing	Sao	2,043	
	Lil	892	
	Belj	1,994	
	Tanda	2,600	These forests, although de-
	Aulas	1,586	marcated, have not been
	Kothi	1,785	reserved.
	Hasn	1,170	
	Maila	2,491	
Total Area		14,561	

Sub-division of the area into blocks or compartments.—The degree of elaboration with which the forest has been sub-divided and the crops described should be briefly explained and justified.

Example.—The working-circle has been sub-divided into twelve blocks, each of an average area of 350 acres and with boundaries which follow either roads, ridges or streams. As the forest is to be worked by the method of selection fellings, and the exploitable trees are growing scattered over this area, a more minute sub-division and description have not been considered necessary.

The boundaries of the compartments are marked by deep blazes on all the boundary trees; whilst, at salient angles, earth mounds with hard wood posts bearing numbered plates, have been placed.

Analysis of the crop.—It should be explained whether the trees have been counted and measured, and whether the areas containing different crops have been differentiated and separately surveyed or have simply been estimated by eye. Where a valuation or enumeration survey has been made, it should be explained whether the stock on the whole area has been counted, or whether the number of trees has been calculated from sample plots. Whatever method of analysis and discription has been employed should be briefly explained and justified.

Example.—No accurate differentiation of the crops in each compartment has been attempted, as this was unnecessary in view of the treatment to be applied. The number of trees has, however, been ascertained by linear surveys run in all directions through every kind of crop. The area so surveyed amounted to 8 per cent. of the area of the working-circle. The results of the enumeration are summarised in the following paragraph, and the detailed figures will be found in Appendix IV.

A brief analysis of the crop, based on any detailed description or on a stock map (if one has been prepared) should be given. Where the number of trees has been counted or estimated, totals for each size, class, kind, etc., should be stated and should be supplemented in an appendix by a detailed record of the survey. If the areas occupied by different crops have been separately examined, the age-classes and the area occupied by each kind of crop should be recorded. Where the stock has been simply assessed by eye, a summary description of the crop in each sub-division may be given.

Example.—The records of the detailed enumeration show that the standing stock may be classed as follows :—

Trees (exploitable) over 2′ in diameter	. . . .	65,871
,, 1½′ to 2′ ,, ,,	. . . .	87,846
,, 1′ to 1½′ ,, ,,	. . . .	63,000
,, below 1′ ,, ,,	. . . .	927,000

Where the crops have been differentiated as to their component age-classes, the summary analysis would be given by area :—

	Acres.
Mature, regular high forest from 100 to 150 years old	1,208
Selection-worked high forest from 40 to 150 years old	1,144
Pole crops from 50 to 90 years old . . .	955
Former coppice, undergoing seed and secondary fellings	87
Young thickets and seedling crops . . .	626
Blanks and glades	167
Total Area	4,187

Purpose with which the forests should be managed.—The object or purpose in view, such as the production of timber of a certain kind and size, the protection of the trees or whatever it may be, should be deduced from the facts recorded in the first part of the Report and be plainly stated.

Example.—The facts recorded in paragraphs 12 to 15 of this Report show that the forests should be worked so as to provide fuel and timber for the local population. Timber of the largest size is not required, and the beams or poles in demand can be obtained from trees of one foot in diameter.

Method of treatment.—The proposals for the management of each working-circle should be discussed separately, commencing with the sylvicultural method of treatment. The method should be explained and its adoption justified by a brief explanation of the reasons which render it advisable or necessary to employ it.

Example.—Although the coppice system introduced in 1871 has not been followed in all its details, its general principles have been adhered to, and, broadly speaking, have given the results that were sought. It is, therefore, proposed to continue this method of coppice with standards. The method of high forest, it might be argued would also furnish both timber and fuel and furnish timber in much larger quantities than would be possible under the method of coppice with standards. The present state of the area, formerly set apart for treatment as high forest, does not, however, favour the conclusion that the method of high forest is suitable to the species which the forest contains. The volume of standing timber is undoubtedly great; but the trees have in general a forced and unhealthy appearance; and it is not improbable that high forest could only be regenerated by artifical means. Every indication points to the conclusion that the principal species thrives best when grown in a state of partial isolation. It finds these conditions when grown as standards over coppice; and, as far as can be seen at present, by adopting this method we are more likely to furnish timber of good quality than by any other method of treatment.

The exploitable age.—The manner in which the exploitable age has been calculated, and the facts on which the calculation has been based, should be stated. As already indicated these facts or considerations relate to the products required, the object with which the forest should be worked, the rate of growth, the prices realised and the net value of the trees standing in the forest.

Example.—It has been stated that the object with which this forest should be worked is to supply fuel and timber for the neighbouring population, and it has been decided that this end may best be accomplished under the coppice method of treatment. It has been shown that the fuel billet required should not exceed 4 inches in diameter, as, if larger, it becomes necessary to split them, and they bring in a lower net price. From the rate of growth of the coppice it is known that this size is attained in about 12 years. It is, therefore, proposed to exploit the forest on a rotation of 12 years. As regards the standards, we have seen that trees of 1 foot in diameter furnish the required small timber, and that they attain this size in about 60 years. It is, therefore, proposed to retain a certain number of the standards of each coupe for five rotations of the coppice.

General scheme of working and calculation of the possibility.—The calculation of the possibility should be based on the analysis of the crops, or, where an enumeration survey has been made, of the standing stock in the forest.

The general scheme of working and the method of calculation having been explained and the possibility determined, the condition of the stock as regards its sufficiency or insufficiency, the arrangement of the age-classes and so forth, should be discussed. The length of the preparatory period, during which it may be necessary to reconstitute the crop or lead it on to normal condition, should also be explained.

Example.—It is proposed to exploit the principal species, teak, by the selection method ; and the length of the felling rotation adopted is 20 years or half the period (40 years) required for a tree of the lowest dimensions of Class II to attain the lowest dimensions of Class I. The valuation surveys show that the age-classes are as follows :—

Class I . . . . 65,871 trees.
„ II . . . . 87,846 „
„ III . . . . 33,000 „
„ IV . . . . 927,000 „

It may be assumed that, as there are 87,846 trees of the 2nd Class, the greater proportion of which will attain exploitable demensions in the course of 40 years. the number of trees which will become exploitable each year is something less than 87,846 ÷ 40 or about 2,000 trees. Under a felling rotation of 20 years, the normal exploitable stock on the ground should therefore be :—

In the area first felled over 20 years ago $\frac{2,000}{20} \times 20$

„ next „ 19 years ago $\frac{2,000}{20} \times 19$

etc., etc., etc.

„ last „ 1 year ago $\frac{2,000}{20} \times 1$.

The normal exploitable stock would, therefore, be $100 \times (20 + 19 + \&c. + 1) = 21,000$ trees.

There is therefore an existing surplus stock of 65,871—21,000 or, say, 44,000 exploitable trees.

The condition of the crop shows that trees of Class III are markedly deficient and that there will be a dearth of exploitable trees 40 years hence, and indeed thereafter until the present stock in Class IV matures 80 years hence.

A preparatory period of 80 years is, therefore, necessary in order to properly constitute the stock; and it is proposed to spread the felling of the surplus stock of 44,000 trees over this period (or rather to diminish the fellings of the younger stock proportionately). The annual fellings should, therefore, be within (44,000÷80) + 2,000 trees, or about 2,550 trees a year.

Period for which the fellings are prescribed.— The time during which the felling operations are prescribed in detail should be explained and justified. In the case of selection fellings the period chosen would be the length of the felling rotation, and the rotation should, therefore, be justified.

Example.—It is desirable, in order that the area worked over each year should not be inconveniently large, that the felling rotation should be as long as possible without risking the loss of trees which die or fall in the interval between the fellings. It is considered that 20 years is the utmost limit that can be allowed under these conditions. The length of time required for trees of the lowest dimension of Class II to attain the lowest exploitable dimension being 40 years, it will be convenient to take that period as the basis in fixing the felling rotation. Consequently a felling rotation of 20 years, or half the period, has been adopted, and operations are prescribed for that length of time.

Areas to be felled annually or periodically.—The formation of the annual or periodic coupes should be explained, and the calculation by which the size of the coupes has been determined should be given with such detail as the circumstances in each case require. Such calculations have already been illustrated by examples.

The allocation of the fellings should be discussed where necessary and the proposals made justified, important departures from the rules regarding the allocation of the fellings being explained. In the case of the selection method, the order followed in making the fellings is of little importance provided that some order is adhered to; but, where one or two species only are exploitable, it may be necessary to form coupes of equal resources as regards these species.

Example.—The order to be followed in making the felligs would, under normal conditions, be as they are numbered 1, 2, 3, etc.; but as coupe No. 14 contains many mature and over-mature trees it is desirable to take it in hand at once. Moreover, as the mature trees are in excess and as there is a surplus stock which must be removed, an excess felling is justifiable. It is, therefore proposed to work coupe No. 14 in addition to coupe No. 1 in the first year, returning again to No. 14 in the usual order in the 14th year of the felling rotation. By this means after the first year, it will be possible to resume the regular order.

Rules regarding the conduct of the fellings.—The special statement of fellings should be accompanied or followed by such general instructions, with regard to the manner in which they should be executed, as may be necessary for the guidance of the local officers. These instructions should not be too elaborate or detailed, but may be drawn up on the supposition that the officers who carry out the provisions of the plan understand their business and will be responsible for the fellings they make.

Example.—The trees are to be felled according to the principles of the method of selection fellings, subject to the following conditions :—
The prescribed number of trees may be felled annually, or the fellings may be carried out, over the whole area to be worked, during three years and in the first or any other year of that period.
No tree under 6 feet in girth at breast-height may be felled, except in certain cases where, in the opinion of the executive officer, the tree shows signs of premature decay.
Isolated trees should not be felled, unless with the object of clearing away cover to help the establishment of seedlings already existing.

It will generally be advantageous to forecast the condition of the crop at the expiry of the period for which the fellings are prescribed. The condition aimed at would of course be in accordance with the purpose with which the forest is worked; and the mere formulation of the anticipated condition would serve to illustrate and explain, better than anything else, the nature of the fellings to be made and the immediate object sought in making them. In many cases the best means of indicating in detail (when detail is necessary) the way in which the fellings should be conducted is by means of suggestions recorded in the " remarks column " of the descriptive statement for compartments or blocks.

In some cases it may be desirable to deal at greater length with the fellings to be made—

Example.—The rest of the periodic block, which contains old fir forest, will be regenerated during the first period by successive fellings, seed, secondary and final. These fellings will be controlled by volume, care being taken to complete, as far as possible, all the seed fellings within the first 12 years of the period. The local officers will execute the regeneration fellings on their own responsibility, in such a way as to secure, with the greatest approach possible to certainty, the natural restocking of the ground. The seed fellings should be made close, the cover being raised by pruning the lower branches, care being bestowed on the condition of the surface-soil. Wherever a thick growth of moss, bilberries or heather covers the ground, it should be removed in wide strips. Should the ground be overrun by herbaceous growth and there is reason to fear that the self-sown seedlings will not thrive, it will be necessary to clean-fell and to re-stock artificially, using the pine as a nurse, and, when the latter is 30 years old, introducing the silver fir and the beech under its shelter. There may chance to be spots where the shelter is sufficient for the direct rearing of the silver fir, and where the soil is of sufficient depth and fertility to render unnecessary the expense of using the pine as a nurse. In this case a middle course should be followed, and pine and silver fir should be sown in alternate strips. In their early years the pines will protect the firs and may afterwards be gradually extracted in accordance with cultural requirements. Saplings and poles that are not vigorous should usually be retained. All those which are too old although yet thriving, that are weedy in appearance, and those in, or which have been under, shade for a long time, ought to disappear. It is only where their presence may be necessary as a protection to younger growth that they should be temporarily retained until the latter has passed the stage at which it requires shelter. When it is considered that a pole crop of silver or spruce fir is in the conditions that render it desirable to preserve it it should be thinned, if necessary, and all old trees which may be considered either useless or prejudicial should be extracted from it.

Tabular statement of fellings.—As a rule the whole of the prescriptions are summed up in a single simple tabular statement containing the following columns:—

> Year or period for which operations are prescribed.
> Area to be taken in hand each year or period.
> Nature of fellings to be made.
> Area or quantity of material to be exploited.
> Remarks.

More than this is seldom required, no matter how complicated the operations prescribed may be. The area to be taken in hand should, if possible, be shown under the different headings: wooded, blank, unproductive. In the last column but one should be entered either the area, the number of trees, or the volume of material to be felled.

Example.—The following example illustrates the adaptation of such a statement to different classes of operations:—

Statement of fellings.

Year or Period	Locality			Area in Acres				Nature of Fellings	Area, Number of Trees, or Volume to be Felled	Remarks
	Forest	Block	No. of coupe or compartment	Wooded	Blank	Unproductive	Total			
Successive regeneration fellings (by volume.)										
1880 to 1889	Lallatun.	2nd. periodic block.	I. II. III.	785	20	3	808	Regeneration.	35,803 cubic feet annually.	The fellings may be preparatory, secondary or final; or all three may be undertaken together by the group method.
Selection fellings (by number of trees.)										
	Anduri	Paimar	I.	271	36	6	313	Selection.	200 trees annually.	Of these, 170 to be sál and 30 sain.
1881-82	Do.	Do.	II.	276	62	10	348			
1882-83	Do.	Do.	III.	250	13	1	264			
...	etc.	etc.	IV.	etc.	etc.	etc.	etc.		etc.	

Statement of fellings—continued.

Year or Period.	Locality.			Area in Acres.				Nature of Fellings.	Area, Number of Trees or Volume to be felled.	Remarks.
	Forest.	Block.	No. of coupe or compartment.	Wooded.	Blank.	Unproductive.	Total.			
Thinnings (by area.)										
1880-81	Anduri	Datwind.	I.	71	22	6	99	Thinnings.	71 acres.	Thinning s over one-tenth of the area (71 acres) to be reposted three times (or once every 10 years) on each area during the period.
1881-82	Do.	Do.	II.	73	2	12	87	Do.	73 ,,	
1882-83	Do.	Latoun	III.	86	32	61	179	Do.	86 ,,	
etc.	etc.	etc.	IV.	etc.	etc.	etc.	etc.	etc.	etc.	
Stored coppice fellings.										
1880-81	Changla	Lodi.	I.	87	6	2	95	Stored coppice.	67 acres.	At each coupe not less than 15 standards are to be reserved per acre on an average.
1881-82	Do.	Do.	II.	92	8	1	101	Do.	92 ,,	
1882-83	Do.	Do.	III.	86	Nil	12	98	Do.	86 ,,	
	etc.	etc.	IV.	etc.	etc.	etc.	etc.	etc.	etc.	

Supplementary provisions.—There are two classes of works of improvement, namely, those which are so connected with the method of treatment that they may be considered as special to the working-circle, and those which are common to the whole area dealt with in the Report.

Thinnings and cleanings, the re-stocking of blanks and the introduction of superior species into the crop obviously belong to the former category; and grazing, although it may be general to the whole area, is nevertheless to some extent dependent on the treatment adopted, at least as regards the closing and opening of blocks to cattle.

Such works or regulations should be dealt with in treating of the working-circle to which they relate. It will be understood that the degree of detail with which these regulations should be drawn up and their nature depend on the circumstances of each case, and that no guide as to what should or should not be provided can here be indicated. Important works of construction should form the subject of a separate report, and should only be mentioned in the plan so far as they affect the management or treatment of the forest.

Improvement fellings.—These should be prescribed by area with such details as regards the manner of conducting them as the circumstances require.

Example.—Improvement fellings should be made at intervals of ten years thus passing twice over the entire area in the course of the felling rotation. The fellings should be made in the following order:—

YEAR.	AREA TO BE OPERATED ON.			REMARKS.
	Block.	Area.	Area to be felled over (estimated).	
1889-90 } 1899-1900 }	Banji	629	167	
1890-91 } 1900-01 }	Sarni	702	206	
1891-92 } 1901-02 }	Rupani	564	221	
1892-93 } 1902-03 }	Chatri	498	196	
1893-94 } 1903-04 }	Masrund	723	205	
1894-95 } 1904-05 }	Kanga	435	189	The estimated cost of these operations R5-8 per acre.
1895-96 } 1905-06 }	Sloh	801	202	
1896-97 } 1906-07 }	Dand	398	186	
1897-98 } 1907-08 }	etc.	etc.	etc.	
1898-99 } 1908-09 }	etc.	etc.	etc.	
TOTAL	......	4,750	1,572	

In the improvement fellings the following work should be done:—
All suppressed deodar seedlings should be relieved from the injurious cover of inferior species, either by the lopping of a branch or two or by the ringing of the immediately over-topping trees.
All deodar trees with crowns contracted on account of the heavy surrounding foliage of other trees, but otherwise in good condition, should be set free by the ringing of some of the latter.
In the vicinity of, or on slopes immediately below, fertile deodar trees, the soil should be prepared for the reception of any seed that may fall by being cleared of all undergrowth and being freed, if necessary, from the thick covering of undecomposed leaves.

In opening out the leaf canopy, it should not be forgotten that deodar in its youth supports a great deal of shade and requires protection, and that bright illumination results in the soil of these forests being overrun with a dense growth of weeds and inferior shrubs. The object of ringing and not felling the obnoxious trees is to uncover gradually the soil and vegetation, to save the heavy outlay that felling would require, and to prevent the ground from being encumbered.

Regulation of rights and concessions.—The plan, in this respect, must carry out the detailed record-of-rights under the forest settlement, if one exists. It should be laid down what areas are to be opened to grazing and for what periods. The number of cattle to be admitted should, where possible, be prescribed ; and generally it should be explained how the rights can be met with the least amount of injury or danger to the forest.

Example.—The annexed table indicates the periods for which the several areas will be opened or closed to grazing. They have, as permitted by the settlement been arranged with a view to giving each area a rest of five years. The number of cattle to be admitted is that given in the statement of rights, and the areas opened will allow 3 acres per head of cattle grazed. When one area is closed, another will be opened ; and, on an average, one-third of the whole area will always be open to grazing. This arrangement in no way contravenes the orders passed by the Forest Settlement Officer.

Years during which opened.	Area opened.		Remarks.
	Name of block.	Area.	
1884 to 1888	Jellaki	3,117	
1889 to 1893	Shahpnr	2,832	
1894 to 1898	Ganchan	2,979	

The Raura block of 106 acres which it is proposed to work regularly will have to be closed. It is free of rights, but grass-cutting can be permitted. The portion traversed by the path from Raura to Mairawana, or about ten acres, which contains chiefly young chir, may, however, be left open to cattle, as it need not be worked and it is inexpedient to close the much-frequented path.

It has already been explained that, as regards rights to timber, it may be necessary either to form the area burdened with the rights into a separate working-circle, in which case the exploitation will be dealt with in the ordinary way as a principal provision of the plan, or to allow for the required material in calculating the possibility. In the latter case the removal of the right-holders' trees would be regulated as was necessary but separately from the fellings made under the principal provisions of the plan.

Example.—The number of trees required annually by the villagers has been estimated as follows from the average consumption of the past five years. These trees will be marked by the Range Officer for removal by the right-holders on production of their passes : —

Description.	Forest from which to be granted.						Remarks.
	Chilisnia.	Karchull.	Padholl.	Dwirsion.	The Chauba.	Total.	
Chir trees, 2' diameter	8	8	12	88	16	132	
,, poles, 1' ,,	60	268	149	76	58	611	
Oaks for charcoal	10	8	12	27	16	73	

Hitherto free-grant trees have been marked in an irregular manner, the same village often obtaining trees from different localities, and the people frequently being allowed to select their own trees regardless of the well-being of the forest. This must cease, and all trees should be marked by the Ranger according to the principles applicable in selection fellings.

Sowings and plantings.—The question of sowings and plantings should be discussed, only such details with regard to their execution being given as appear to be necessary for the guidance of the local officers.

Example.—It is proposed to introduce tun and other superior kinds of trees into the crop by planting in the coppice. The want of good species to serve as standards and the increased value that such standards would add to the forest have been fully explained. To enable these plants to hold their own in the dense coppice growth (only certain kinds of shade-supporting species could be so introduced) good-sized seedlings should be planted in pits. Nurseries should be established in the compartments to be felled three years in advance of the felling.

Sites for nurseries should be chosen in well-drained localities. The nursery beds should be terraced, and the seeds should be sown in lines, a foot apart, in November and December. Whilst in the nursery, the young plants should be protected in seasons of drought and frost by grass tatties raised a few feet off the ground. The young seedlings should be put out immediately the rains set in.

Roads, buildings, and other works.—The improvements indicated above are more or less connected with the method of treatment adopted. There are, however, many works, such as the improvement of boundaries, the construction of roads, and buildings, and, in many cases, the clearing of fire-lines which may concern all the working-circles.

The estimated cost of such works as may be proposed should be given in the report. The application of a plan may necessiate the construction of very considerable export works and roads, but such undertakings should form the subject of separate reports and should only be briefly referred to in the working plan.

Example : Boundaries.—The forests have been merely temporarily demarcated by *kutcha* pillars many of which have already fallen down. In some cases it will be possible to treat collectively as one block several at present separately demarcated for instance, Sikri, Rupani, and Chatri; and also Kalwara, Sani, Padri, Bohar, Chirindi and Rumbo. I nt where this is inexpedient the forests should be re-demarcated by *pucka* boundary pillars It is important that new pillars should be of the best possible description. Forest records of former years show that demarcation work has frequently had to be done time after time owing to the cheap and unstable nature of the materials used. It is therefore proposed to construct of the most suitable stone (generally slate) found in the locality, solid masonry pillars about 2 feet square at base, 1 foot 6 inches square at top, and 2 feet 6 inches high, on a solid foundation 3 feet square. It is estimated that each pillar of this description will cost from R 2 to R 4 according to its position. Owing to the configuration of the ground, one pillar is frequently not visible from the next, and there is often, in consequence, uncertainty as to how the boundary runs. The remedy is to cut a two-foot path or line through the forest from pillar to pillar, wherever the boundary is not a natural one or a path does not already exist. In paragraph 6 it is estimated that there are 73 miles of artificial boundary; so that, if 8 miles were made every year, the whole work would be completed in about ten years.

Fire-protection.—As stated in Part I, it is accepted that unless forests open to rights are fire-protected their deterioration must continue, and they must in course of time, disappear in accessible places, especially if the demand of right-holders for timber and fuel is supplied, as in the past, from such localities only. It is therefore proposed to extend protection to the following blocks :—

Forests.	Area in acres.	Length of fire-line required in miles.
Sukrau	11,132	11
Gwalgarh	13,435	36
Rawasan	15,933	59
Total	40,500	106

All the other forests in the area dealt with, viz., Jogi Chur, Ander-Majhera, and Kauria Chaur are already protected. The Buriwala-Mabera block is an island, and requires no fire-line for its protection; but grazing might be stopped in it at the beginning of the fire season. The extra cost which this protection will involve is estimated, on the cost of similiar protection in other areas in the division, at Rs. 450 a year. This will bring the total cost of fire-protection in the tract to Rs. 900 annually.

Communications and buildings.—The roads, bridges, and buildings enumerated in paragraphs 80 to 85, except the sleeper-carrying paths in the blocks which have been worked out, should be kept in repair. A year before the selection fellings in the Deota forest are brought to a conclusion it will be necessary to make a good bridle-path by widening the present path from Deota; and a similiar path should be made later to Bamsu from Deota to Sahlra along the main ridge, a short one joining it from Bamsu. Roads, passable for pack animals carrying food for the sawyers and coolies, must also be made along the right bank of the Tons from Tadiar, passing by the mouth of the Bamsugad to that of the Kunigad, down which will come all timber from the Sahlra forest ; whilst branch roads will be required up the two gads referred to and also one from Naintwar bridge to Koarbo, the road on the left bank of the Tons as far as that bridge being widened and improved.

Forest rest-houses and servants' quarters should be built at the camping ground in the Sahlra forests and at the mouth of the Kunigad. The Bamsu hut must also be enlarged and improved. Houses for the Forest Ranger and chaukis for Forest Guards will be required at Ba su, Sahlra Kunigad and Naintwar. Godowns for the food-supply are also necessary at Bamsu, Sahlra and Naintwar. The cost of these various works will, on an average, amount to R 5,000 a year.

Summary of works of improvement.—A summary of the works of improvement may, in some cases, be given with advantage. A brief description of the works, the probable date of their execution as well as of their cost, should be indicated.

Example.—Summary of works of improvement.

			Rs.	
1892-93	Joli	Facilitating the reproduction of deodar, as described, in about 280 acres.	560	Includes cost of fencing small portions where necessary.
	Padri, Bohar	Collecting seed, preparing and sowing 20,000 square feet of nurseries, as described.	165	

			₨	
1892-93	Padri, Bohar	Planting 20 acres, as described.	260	
	Ditto	Resowing old nurseries : 20,000 square feet.	40	
	Joli Thali	Opening out a path through the forests, and from the forests to Sao river, to join path along that river : about 3 miles.	300	
	Sao	Improving the village road along the Sao nalla : about 3 miles:	250	
	Sai	Building a small forest rest-house on road leading to forest in the Sao valley.	500	Not including value of the timber.
	Joli-Thali	Demarcating the forest boundaries with about 30 *pucca* pillars, as described.	60	

Miscellaneous provisions.—The working-plan may fittingly prescribe the up-keep of records, the conduct of experiments with a view to the future revision of the plan on more accurate *data*, or for other purposes.

Example.—A journal, as prescribed under Section 89 of the Forest Code, must be kept up. This book will contain, separately for each block, a register of the operation carried out therein, the yield from the fellings, thinnings, dead and wind-fallen trees, etc., etc., etc. In addition to this, a portion of the journal should be devoted to a general summary of each year's work, to notes on experiments and observations made, cost of exploitation, revenue, and expenditure ; and so forth for the forests *as a whole*.

Measurements of the rate of growth of the oak should be made regularly once a year in the sample plot and should be entered in the journal. The examination of the concentric rings of growth in conifers should be continued. The record should be kept by compartments, and both the aspect and altitude should be invariably noted.

Changes in the forest staff.—Any alterations proposed should be indicated with such detail as circumstances require, and should be justified by reference to facts recorded in the first part of the Report.

Example.—It has been already fully explained that the staff is inadequate and must be increased if the present proposals for working the forest are accepted.

The following protective and executive establishment will be required. A statement showing how it is proposed to distribute and employ these men is attached to this report. Even with the proposed increase, the average forest area (formed of a large number of small forests scattered over rugged country) under each ranger will be 17,000 acres, and under each guard 1,500 acres—

		₨				₨	
Upper Ravi forests : four ranges.	3 Forest Rangers on	50	to	150 a month	.	150 monthly.	
	5 Foresters on	15	„	40	„	. 115	„
	45 Guards on	5	„	10	„	. 285	„
Lower Ravi Range.	1 Forester on	50	„	150	„	. 50	„
	2 Foresters on	15	„	40	„	. 45	„
	22 Guards on	5	„	10	„	. 135	„
	Total estimated cost at starting					780	
	Present sanctioned monthly expenditure					610	
	Proposed monthly increase					170	

L

This does not include the establishment employed in the forest round Dalhousie which will be separately reported on.

Forecast of financial results.—The report should conclude with a forecast of the anticipated financial results under the management proposed. The forecast should be criticised and compared with the results obtained in past years.

Example.—It is impossible to do more than estimate the *average annual* revenue. It will be noticed, from the following statement of the unticipated revenue and expenditure during the next 10 years, that but little profit from a financial point of view will be obtained from the working of the forests. It is not intended that the position should be otherwise. The object Government holds in view in working these forests is, not to trade as timber speculators, but to secure a regular supply of timber for the neighbouring town, and to manage the forests so that they shall be always capable of furnishing a sufficient supply. It is, therefore, here proposed to devote to the improvement of the forests, which are in a dangerous condition as regards their future, the greater portion of the anticipated net revenue from the timber transactions. The anticipated receipts during the next 10 years compare, it may be thought, too favourably with the actual results of the past decade. It will be seen, however, that it is proposed to fell more than double the past average annual number of trees.

Sources of revenue and heads of expenditure.	Estimated average annual receipts and charges.	Average annual receipts and charges during past 10 years.	REMARKS.
	R	R	
Timber and fuel	p2,000	...	
Minor produce and miscellaneous	1,300	...	
TOTAL RECEIPTS	53,300	29,278	
Timber works	11,000*	5,200	* This includes one-seventh of the total proposed expenditure which will be incurred on the purchase of new tramway stock.
Roads and buildings	6,000	2,176	
Plantations and other works of improvement	21,000	8,121	
Salaries of establishment, including four-seventh of the divisional controlling and office establishment charges	8,500	6,232	
TOTAL EXPENDITURE	46,500	21,729	
NET SURPLUS	6,800	7,549	

APPENDICES.

Maps.—The working-plan should be supplemented by such maps as are required to clearly illustrate the general position of the forests as well as their proposed exploitation and management. Ordinarily the maps would include, especially if the total area concerned is very considerable—

   (1) A general map on a small scale showing the whole tract dealt with, the distribution of the

different forests, the boundaries of working circles and of administrative charges, etc., etc.

(2) Separate maps for each circle, on a scale not less than 4 inches=1 mile, indicating the boundaries of blocks, compartments and coupes.

Description of the crop in each block.—It is generally expedient, when a plan of a permanent and detailed character is prepared, to record separately for each sub-division into which the forest has been divided, and with such minuteness as the circumstances of each case require, the more important points connected with—

> (a) the *situation*, relative position of the area and (in hilly country) aspect and slope of the ground;
> (b) the *soil*, nature of underlying rock, state of the surface soil, its composition and physical condition, depth and general fertility;
> (c) *the composition and condition of the standing crop*, type or class of forest; component species and their relative proportions, age, density, state of growth including reproduction; past treatment, most suitable treatment; general remarks.

It has already been explained that much of this information may be graphically represented in a *stock map*, and that such maps may supplement, if they do not replace, the written descriptions. In any case the written record should, in order to fulfil its purpose, be as brief as is consistent with clearness. Lengthy descriptions defeat the object with which the record is framed, as the mind fails to grasp the picture offered and loses itself in details. With the object of rendering these descriptions as little cumbersome as possible they are usually made in the form of a tabular statement such as that reproduced below. The growing stock must be described with accuracy. In describing a coppice, the coppice and the over-wood or standards should be taken account of separately. In the case of crops for which the possibility has been prescribed by area only, no enumeration of standing stock would be made. The following are some examples:—

148

Name of block.	Compartment or coupe.	Area in acres. Wooded.	Area in acres. Blank.	Area in acres. Unproductive.	Total.	Situation and soil.	Condition of the crop.	Standing Stock. Description.	Standing Stock. Number of acre.	Remarks.
Datmir	...	103	26	2	131	*General description.—* Bed of nullah, gentle slope facing the west. Soil deep, resting on shale, rich and fertile.	*For area fellings.* On both sides of nullah, mature seedling sal forest, more or less pure or mixed with sain ; vigorous but with no similar scrub jungle. Rest of area, irregular scrub jungle with a dense growth of bushes and many sal seedlings interspersed among them. Numerous small blanks. No fellings have been made, but area has been constantly grazed over and burnt.	Sál more or less pure. Sál with sain and mixed species. Scrub Grass blanks Unproductive	Acres. 31 11 60 26 2	No fellings of sal should be made until reproduction improved by exclusion of grazing.
Sagli	...	533	21	8	562	*General description.—* West flank of spur; aspect westerly, but northerly and easterly on side spurs; gradients steep, often precipitous; elevation 6,500 to 8,000. Soil rich loam well covered, generally deep but shallow at summit where rock outcrops.	*Selection treatment.* In upper portion kharsu oak with blue pine, etc., deodar rare. The deodar increases on descending, but is mixed with spruce and silver fir. All ages represented, but trees mostly mature ; many over-mature. Density varying ; reproduction fair ; scattered seedling of spruce and silver fir over patches in all lower portions; small blanks here and there throughout. Selection fellings made in more accessible positions only.	Deodar exploitable. Deodar 1½' to 2' diameter. Do, 1½' to 1' diameter. Unsound but exploitable deodar.	No. 1,090 627 606 396	Selection felling to remove over-mature deodar urgently required.
Dhasri	...	1	87	3	1	91	*General description.—Coppice with standards.* Saddle back on ridge. Rocky, poor soil.	*Coppice*, irregular of oak with numerous blanks. *Standards*, pine usually of one rotation, but several old over-mature trees.	Standards of one rotation. Do. old over-mature	No. 365 68	The removal of the old over-mature standards and the regularisation of the coppice and restocking of blank area, necessary.
Bandla	...	6	207	6	...	213	*General description.—Method of successive regeneration fellings.* Occupies lower basin of nullah; aspects northerly and occasionally westerly. Underlying rock shales; soil loose sandy clay.	Mature seedling forest of nearly pure deodar, 130 to 160 years; complete, vigorous and well growing. Reproduction good; groups of thickets 18 to 20 years; one large blank of 6 acres.	Deodar over 2' diameter. Do., 1½' to 2' Do., 1' to 1½' Do., ½' to 1' Do. unsound 2' Other species over 2' Other species under 2'.	cub. ft. 3,60,090 42,600 78,900 1,85,700 30,000 60,200 80,100	Regeneration fellings immediately required.

Record of results of valuation surveys.—In the following example of a record of valuation surveys it is assumed that there are three distinct types of forest, and that it is advisable to deal with each type separately and to take sample plots in each :—

Name of Forest or Block	Type of forest	Area in acres	Sample plot	Area of sample plot in acres	Trees counted on Plots								Total trees in compartment or type								Remarks
					Deodar				Spruce				Deodar				Spruce				
					I	II	III	IV	I	II	III	IV	I	II	III	IV	I	II	III	IV	
Bagti	(a)	80	I	1·5	82	105	209	502	...	63	126	280	} 1,880	3,740	6,300	14,830	...	...	...	...	The *types of forest* are:— (a) Deodar, almost pure, with a few oaks and other broad-leaved species.
			II	1·5	27	83	108	124	...	105	169	452									
	(b)	120	I	1·5	6	27	46	100	53	107	230	400	} 840	1,770	2,230	5,400	2,040	4,410	9,190	17,700	(b) Mixed forest of deodar, spruce and silver firs, with broad leaved species.
			II	2·5	13	33	28	50	16	40	86	130									
	(c)	60	I	2·0	...	...	...	...	27	48	160	260	...	...	...	...	810	1,440	4,900	8,400	(c) Mixed forest of spruce and broad-leaved species, without deodar.
Total number of trees in block													2,120	5,510	8,530	19,230	2,850	5,850	13,990	26,100	The *sizes* enumerated are, in girth:— I.—Above 6' II.—4' 6"–6' III.—3'–4' 6" IV.—18"–3'

Record of rate of growth.—This should show the size of the trees experimented on and the *mean radius* (if rings are counted on stumps) or the *mean depth bored* (if Pressler's gauge is used), and the average number of rings per inch of radius. It is usual to group the trees in diameter classes and to calculate the average rate of growth for each class; as this shows at once the number of years required for the trees of one class to pass into another. A form of record, such as the following, may be used :—

Class or diameter of tree.	Average radius or depth bored.	Number of rings counted.	Average number of rings per inch of radius.	Remarks.
	Pressler's gauge.			
Below 12 inches diameter.	2·1	26		
	2·0	21	11·4	In moist valley.
	2·3	36		
	1·5	20	11·7	On high ridge, but good soil.
	1·8	26		
	2·0	24	13·2	Valley moist.
	Stump	*countings.*		
	5·2	78	15·0	Stiff loam soil, valley.
	5·1	51	10·0	Limestone soil, valley.
	4·6	74	16·1	On exposed ridge, but fair shelter.
	Average for class	. .	13	

CHAPTER V.—CONCLUDING REMARKS.

ENSURING THE PERMANENCY OF THE PLAN.

Laying out the coupes.—When the provisions of a working-plan have been drawn up there still remains, as a rule, the marking off on the ground of the sites of the annual fellings or the coupes. Until this has been done the plan is incomplete. It may be compared to a forest delineated on a map but not marked on the ground by material signs.

The degree of detail with which the coupes should be marked off depends on the circumstances of each case. In working-circles treated by the selection method or subjected to restoration fellings, for instance, each coupe will, as a rule, be formed of one or more blocks limited by natural or artificial land-marks. In such cases no further demarcation obviously is required. It sometimes happens, however, that the limits of the blocks are roads or fire-lines, the construction of which forms part of the provisions of the plan itself. When possible, therefore, it is well to lay out the roads or fire-lines on the ground before finishing the plan. This course will tend to ensure its provisions being adhered to and the sub-divisions it deals with being preserved, and is in every way advantageous. A well-considered arrangement of fire-traces and of roads for the extraction of the produce is of the first importance ; but if the works are only verbally indicated there is every chance of the directions being lost sight of. In coppice treatment especially, where the coupes are comparatively small, the marking-off of the roads or rides is of much importance and very largely increases the value both of the plan and of the forest.

THE CONTROL OF WORKING-PLANS.*

Necessity for simplicity in the control.—Working-plans are of no use if they are not adhered to, and proper adherence can only be secured by an effective control. Much of their

---

* The remarks under this head are an expression of the writer's personal views. The control exercised for forests under the Government of India is detailed in Chapter II of the Forest Department Code.

value will also be lost unless a careful record is kept of the works carried out and of the financial results secured. A sound system of control is therefore necessary; but in order that the control may be effective it should be of a very simple nature requiring only a few entries, once a year or so, in the records regarding each forest or working-circle. For the same reason the control should be of such a general nature that an uniform kind of record can be used for all forests and for all systems of working.

It is unfortunate that the truth of these propositions has not been altogether realised in India. The control forms first introduced were impracticable; because the anticipated outturn of each felling was entered as a provision of the plan. It is not too much to say that the attempt thus made to keep up the control forms gave more trouble than the information recorded was worth.

Nature of control required.—The facts to be recorded for purposes of control may be classed as follows under three distinct heads :—

(1) The work done or fellings made as compared with the provisions of the plan.
(2) The gross "yield" and the "outturn."
(3) Financial results.

Controlling the provisions of the plan.—The control can be effected in a very simple manner. The plan amounts (*a*) to the prescription of certain sylvicultural rules to be applied over definite areas, with the addition, in some cases, of a limitation as regards the quantity of material to be felled; and (*b*) to the prescription of certain works of improvement. All that it is necessary to record and to control can, therefore, be brought under the following headings :—

(*a*) Year or period of operation prescribed.
(*b*) Locality to be exploited or in which the prescribed work is to be carried out.
(*c*) Nature of felling or other cultural operation to be made or of work of improvement to be undertaken.
(*d*) Quantity of material or number of acres to be exploited, or expenditure to be incurred on a prescribed work.

The work done, in accordance with the plan or otherwise, can therefore be checked from a record such as the following which need not usually be prepared oftener than once yearly.

Control Book, Bunsi Range, Puri Division.

Year.	PROVISIONS OF WORKING PLAN.				OPERATIONS CARRIED OUT.			
	Nature of felling or other operation to be carried out.	Locality.		Quantity of material, number of acres to be exploited, or expenditure to be incurred.	Nature of felling or other operation carried out.	Locality.		Quantity of material, area exploited, or expenditure incurred.
		Name of block.	Area.			Name of block.	Area in acres.	
1890-91	Selection felling.	Bansi, Maila		*Derbunga Circle.* 250, 370	650 chir trees.	Selection fellings.	Bansi, Maila (half)	435	450 chir trees.
	Artificial restocking with chir.	Sadkot	150	R750	Nothing done.			
1890-91	Thinnings	Loha	350	*Uparkot Circle.* 70 acres	Thinnings	Loha	350	70 acres.
1890-91	Clearing firelines.	...	...	*All Circles.* 10 miles. R350	Clearing firelines.	...	...	7 miles, R210.

From this form it will be seen that works of improvement, such as climber-cutting, thinnings, etc., can be entered as well as other operations, such as fellings and the construction of roads and houses. As a rule the expenditure is the best means of controlling works of improvement. Thus, where a certain forest requires roads in order to develop it and to increase the net price of the produce sold from it, it would be sufficient to prescribe the annual expenditure on roads of a certain sum, or a certain percentage of the net revenue after deducting the expenditure on timber works, until the net price was in this way sufficiently raised. Such a prescription could be far more readily and usefully controlled than a detailed statement of definite roads or buildings to be constructed, though such works might usefully be indicated in the plan without being prescribed. Until working-plans are drawn up and controlled with simplicity, as indicated above, they will give rise to useless correspondence and will not be adhered to.

Recording the results of working—Little need be said with regard to the record of outturn, except that it should be quite separate from the statement of the provisions of the plan proper.

Record of financial results.—The record of financial results should be nothing more than a brief abstract of the annual receipts and expenditure. It is unnecessary to furnish details; but it is well, in order that the capital invested in the forest may be known, and its amount tested from time to time, to distinguish *capital expenditure* sunk in the forest from *current outlay* on annually recurring works of protection, on repairs, on protective establishment, and so forth.

A form, such as the following, could be written up annually with little trouble and would record all the necessary information both as regards the capital expended and the financial results of working for the year:—

	RECEIPTS.			EXPENDITURE.				
YEAR.	Principal produce.	Minor produce.	Total receipts	On timber works.	Capital outlay on works of improvement.	Current outlay on protection and maintenance, including establishment.	Total expenditure.	Net money yield per acre.
	R	R	R	R	R	R	R	R
1890-91	27,032	2,001	29,033	*Derbunga Circle.* 3,705	7,629	10,223	21,557	...
	...	16,509	16,509	*Uparkot Circle.* Nil.	309	5,902	6,211	...

Areas for which separate control statements should be prepared.—It is a question of some importance whether the records alluded to above should be kept up separately for each working-circle or merely for each locality or administrative charge. There can be no doubt that it would be exceedingly useful if the forms were prepared for each working circle, and that without such a record much of the information, especially regarding the yield and financial results, is valueless. Also that, as regards the fellings and other main provisions of a working-plan, a separate record and comparison for each working-circle is indispensable.

But the weakness of the forest staff, the nature of the work done, and the manner in which such records must be kept up often render it impossible to prepare separate records for each working-circle. It should also be borne in mind that many works, such as buildings, roads, fire-lines, etc., are not always connected with any one particular circle, but may be common to the whole area concerned. It is therefore, necessary, if the returns are to be prepared without an undue expenditure of time and labour, that considerable latitude should be allowed in their up-keep; and it will be best to require their submission only for each separate locality or executive charge, such as a range, the working circles being distinguished from one another by writing the name of each across the return or statement as in the examples given above. This matter, although one of detail, is of considerable practical importance.

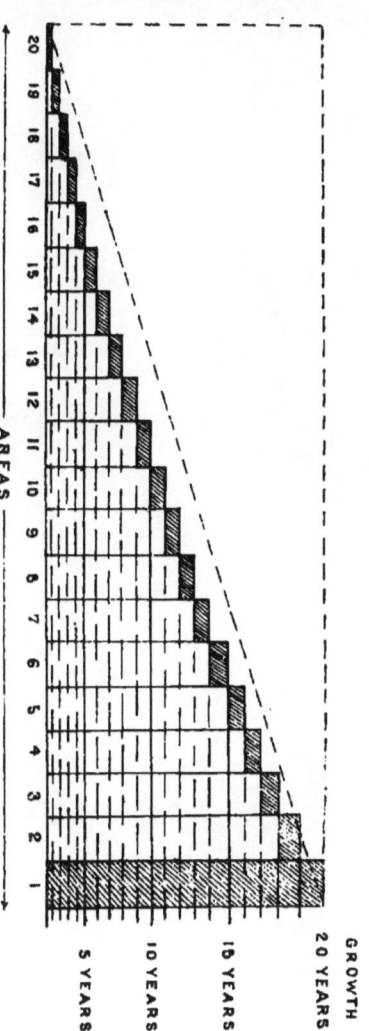

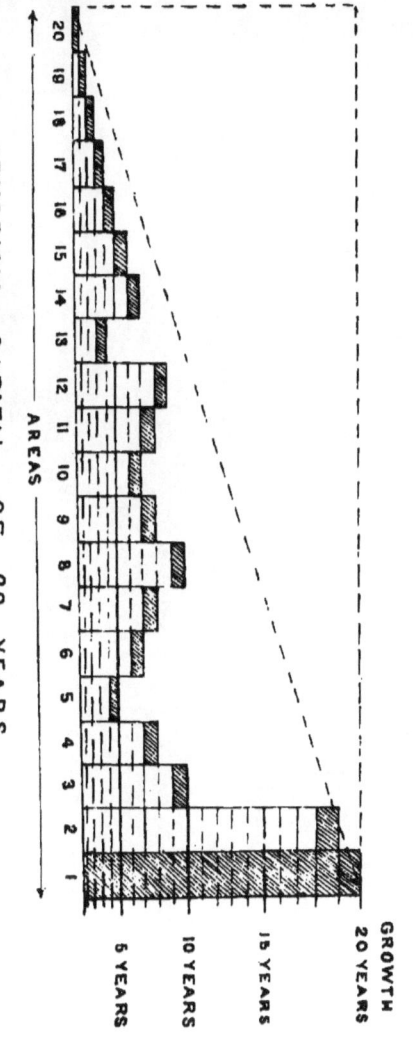

ABNORMAL CAPITAL OF 20 YEARS
FIG 2.

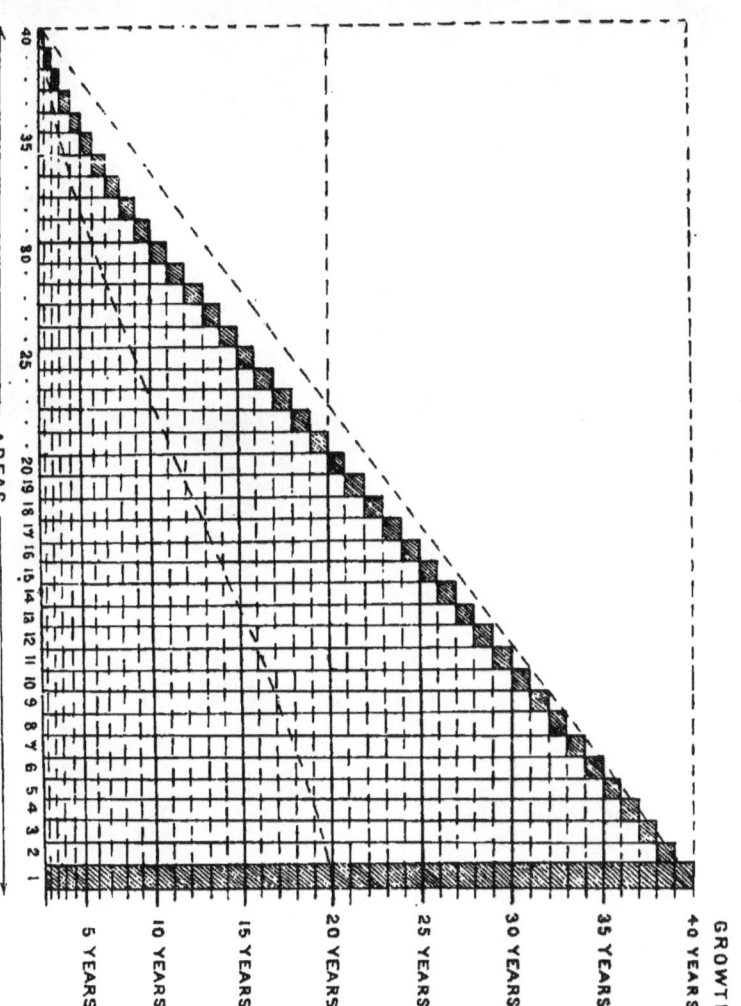

NORMAL CAPITAL OF 40 YEARS.

FIG. 3.

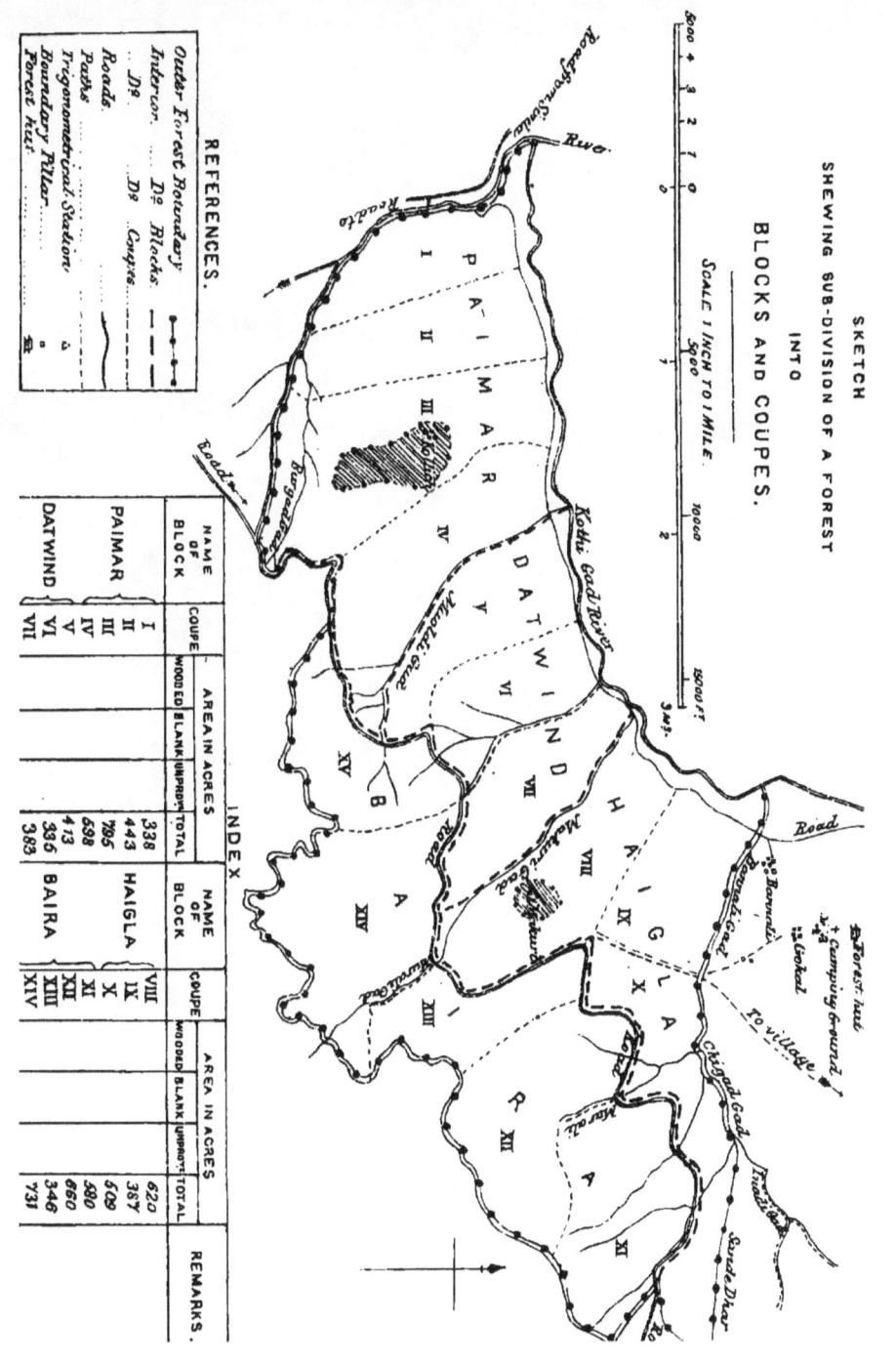

www.ingramcontent.com/pod-product-compliance
Lightning Source LLC
Chambersburg PA
CBHW022112160426
43197CB00009B/990